Lecture Notes on CALCULUS at POSTECH

for students who learned some Calculus before

Kang-Tae Kim

KYOWOO

Kang-Tae Kim
Department of Mathematics
Pohang University of Science and Technology (POSTECH)
Pohang City, GyeongSangBukDo 37673
The Republic of Korea

Kyowoo Co., Ltd.
8, Yangnyeongsi-ro, Dongdaemun-gu,
Seoul, Korea 130-861,

Phone: 82-2-925-2861
F a x: 82-2-925-2860
e-mail: kyowoo@kyowoo.co.kr
http://www.kyowoo.co.kr

Printed in the Republic of Korea.
ISBN 979-11-251-0314-1 (93410)

₩ 17,000

Preface to Part I

Most POSTECH freshmen start their mathematical study with Calculus. But the current high school students entering POSTECH have learned much of the technical part of the first year college level calculus already.

It seems, generally, beneficial that the student learn some advanced concepts and methods early. The downside is that they learn various techniques for problem solving without really appreciating the depth of knowledge inert in Calculus. I have seen so many cases that the student "know" how to calculate the equation of tangent line, but he/she does not even know what the definition of tangent line precisely is!

We, professors of mathematics, were alarmed further to learn that, Korean high school mathematics books of these days teach the conclusion of the fundamental theorem of calculus as the definition of definite integral! While in shock, we asked,

> "Then, how can they know the definite integral actually computes the area under the graph of the given function?"

All we received in return were sheepish smirks.

When we try to teach Calculus again in standard ways, in particular, following the big thousand-pages-long textbooks (mostly from the U. S.), many POSTECH freshmen find it boring and tiring, because they have learned it already. But then, when we try to discuss something from the same books which are not necessarily deeper but they have not seen before, many of them simply give up.

What shall we do? While we have not reached at any effective solutions, we decided to take an action along a nonstandard approach. Trusting that POSTECH freshmen are well-prepared and equipped with good problem solving skills, we start the course with techniques of integration, which are not far from what they handled at the final stage before entering POSTECH. Then we jump into the power series, which will enrich the list of functions that will be used throughout their lives as science and engineering majors.

We then introduce the Taylor expansion and series as well as the estimation of the remainder term. This will be the first half of the first semester calculus.

After the midsemester, we go into relatively simple ordinary differential equations. There the student will see the "use" of what they learned in the first half. They can taste also the beauty of calculus through those ordinary differential equations and some of their applications.

Up to this stage, I hope that the student sees gradually that their previous knowledge is also an important part of Calculus and yet, it was on rather not-so-firm a foundation. Therefore, in Chapter 5, we give a taste of proper calculus with so-called "ϵ and δ." Here we try to provide the correct viewpoint (and understanding, if we may hope for) towards Calculus proper. I believe that the core of Calculus is in this chapter, and the contents of all other chapters can be viewed as applications!

Then we come back to the application again introducing the elementary transcendental functions.

The 2019-01 semester consists of only 14 weeks of teaching which consists of 28 class meetings of 75 minutes each. So we decided to mark many sections as "optional," even though they are at least as important as any other topics.

This could be a wild attempt, I know. But I hope that POSTECH freshmen of 2019 find it worth while, because pedagogical success in POSTECH was in the center of my motivation for preparing these lecture notes. This book has no exercise problem sections; I had the instructors as well as the teaching assistants provide them (about 10 problems each week).

Of course, these whole lecture notes are based upon what I learned from many great teachers, to whom I tribute all the due credits.

K.-T. K.

2019 @ POSTECH

Preface to Part II

Part II was originally written as a separate book; thus are these notes on Calculus I and II as well as this preface 2, originally the preface of that volume, composed as such.

Part II is for Calculus of real-valued functions with 2 or more variables, i.e., the multidimensional Calculus. As Calculus usually means differentiation, integration, and their immediate applications, these notes stay well within such limitations. On the other hand we shall give more emphasis on the theoretical part, leaving the applications to the hands of the instructor and the student.

Along the exposition of these notes, the student will start with Chapter 7 handling the basic (linear) geometry, such as vectors, lines, planes, angle, length, vector equations and parametric equations that represent those linear objects in the Euclidean spaces. Although the exposition can be applied to all dimensions, we focus more on 2 and 3 dimensional cases. These are necessary, as Calculus leads us to understand the functions via differentiation, reducing the problems to the analysis and geometry of linear objects.

Then we discuss the differentiation of functions such as $z = f(x_1, \dots, x_n)$. The differentiability of such a function is of course (as it should be!) the existence of the tangent hyperplane at the point in consideration. Such development is strictly analogous (and therefore is the orthodox one!) to the differentiability of the functions in one variable, introduced in Chapters 1-6 (the definitions are precisely presented in Chapter 5), using the correct definition of the limit, so-called "ϵ-and-δ" in Chapter 5. Then, from these thoughts, the concept of partial derivatives spring out naturally. From this point on, commonly discussed concepts such as gradients, directional derivatives and so forth receive natural exposition with analytic and geometric understanding.

Then we discuss the traditional min-max principle, a. k. a. the optimization. We do this via the Taylor expansion of multivariable functions, which is of course a variation of The Mean Value Theorem. In addition, the brilliant method called "Lagrange multipliers" is also discussed.

Then these notes lead the student into some basics to the application of differential calculus. The POSTECH students have, over time, appreciated it when the instructors mention the idea of methods such as linear programming, regression line, graphing the surfaces in 3 dimensional Euclidean spaces (via the method of level curves) from the equations such as $F(x, y, z) = 0$. For the integral calculus which will come up shortly after learning the contents of these notes, I also included discussions on how to handle the curves in the xy-plane given by equation in polar coordinates, and on curves on the surfaces prescribed.

Then the midterm examination week of the second semester came and passed.

As the semester resumed, the instruction resumed with multidimensional integrals and integral theorems. The contents are in Chapters 11-13. Integrals on domains in $\mathbb{R}^n$ as well as on lines (in fact, curves) and surfaces are handled.

Then for the study of these multidimensional integrals, I decided to dare, introducing the use of *wedge products*, *differential forms*, *exterior differentiation* and *pullback operations.* This, to my pleasant surprise, was well accepted by POSTECH students, even though I treated these concepts as part of the well-established formalism for the integrals with "orientation" which are the subject we learn at this stage of multidimensional Calculus. Of course there is always a difficulty and danger with such teaching. But students understood my point when I mentioned that these are presented without full expositions and justifications, as many other topics in Calculus courses are.

The exposition of this volume is obviously not self-contained. It was never intended to be, as it is lecture notes. I hope that the instructor will find it useful in organizing their own lectures on Calculus.

K.-T. K.

Grading policy and ... ,

As the instruction progressed, 10 problems have been assigned each week as homework. The selection of the problems was an important part of the duties of Teaching Assistants. They also ran the recitation session of 50 minutes every week, where they discussed how to solve the assigned problems. Homework was neither collected nor graded.

But it was announced to the class through the written syllabus that at least 70% of each examination will be organized on and from, the level and the contents, of the homework problems. And the promise was kept.

This strongly encouraged the student to study the homework problems and to ask essential questions to TAs. At the end of each semester, the student would have studied approximately 150 problems which explained the contents of the course. Also in the second semester, exams sometimes included problem(s) from the earlier exams. This way recommended students to learn the problems of the past examinations.

Keeping the promise was one thing. But it was more important for the student realize that their earnest effort following the class policy pays fairly. The grading scale was an absolute one; 90 or better receives A+, 80 or better A0, 70 A-, 60 B+, 50 B0, 40 or better receives B-. Below 40 had no particular announcements in the beginning of the semester, but the student understood that those lower achievements will be treated according to the general policy. The grades were understood as absolutely fair by the students, even though the grading essentially did not allow partial credits.

The calculus class of 2019 at POSTECH scored better than 65/100 on average, with the exam problems included in these lecture notes. Overall, more than 35% students received A- or better as their final grades. But more importantly to the instructors, students told us that they really learned Calculus throughout the year 2019.

Contents

Part I

Calculus in 1 dimension

Course Plan for Semester 2019-01

Weeks 1 & 2: Techniques of integration (Chapter 1)

Weeks 3, 4: Chapter 2. Functions by power series
Section 2.3.4 is Optional. The range and schedule for the Exam 1 will be announced.

Weeks 5, 6 & 7: Chapter 3. Taylor development

Week 8: Midterm Examination week. (No class, except for the exams)

Weeks 9 & 10: Chapter 4. Some differential equations

Weeks 11-13: Chapter 5. Calculus proper
Exam 3 will be sometime in the 12th week; schedule is to be announced.
5.9-5.10 will be taught rigorously but without proofs or computations.

Weeks 14 & 15: Chapter 6. Basic Transcendental functions

Week 16: Final examination week.

Chapter 1

Techniques of Integration

We start with some (applications) of Calculus from students' previous learning. They should have learned how to differentiate basic functions such as:

(1) $\frac{d}{dx}(x^r) = (x^r)' = rx^{r-1}$ for any real number r.

(2) $(e^x)' = e^x$, where e is Euler's number (also known as Napier's constant) which satisfies $e = \lim_{n\to\infty} \left(1 + \frac{1}{n}\right)^n$ etc.

(3) $(\ln|x|)' = \frac{1}{x}$ for $x \neq 0$.

(4) $(\sin x)' = \cos x, (\cos x)' = -\sin x$.

Students may have also learned the property of differentiation such as

(D1) $(f(x) + g(x))' = f'(x) + g'(x)$.

(D2) $(cf(x))' = cf'(x)$ for any constant c.

(D3) *Leibniz's rule*: $(f(x)g(x))' = f'(x)g(x) + f(x)g'(x)$

Furthermore, the indefinite integral of a function $f(x)$ is a function whose derivative coincides with $f(x)$. The indefinite integral, denoted conventionally by $\int f(x)dx$, is a function of variable x satisfying

$$\frac{d}{dx}\Big(\int f(x)dx\Big) = f(x).$$

Students also learned definite integral (what is it?) and they should know(!) how to evaluate them when the integrands are concrete functions.

The computations are based upon the rules above (but why are they true?) and the rule for differentiation of composite functions, so-called the *Chain rule* which is

$$(f(g(x)))' = f'(g(x)) \cdot g'(x)$$

if everything works out nicely. (Again, the proof?) This will even allow students to differentiate the more abstract functions such as the inverse sine function as well as further *inverse trigonometric functions* etc.

These lecture notes will lead students to learn all these concepts and techniques conceptually and precisely (as much as possible!) as the exposition progresses, but in this chapter, we shall deal with technical computations for the examples arising in Calculus of functions in one variable.

1.1 Integration by parts

Theorem 1.1.1

$$\int udv = \int u(x)v'(x)dx = u(x)v(x) - \int v(x)u'(x)dx = uv - \int vdu.$$

Proof. Immediate from

$$\frac{d}{dx}(uv) = u\frac{dv}{dx} + v\frac{du}{dx}.$$

□

Example 1.1.1 Find an explicit form of $\int xe^x dx$.

Solution.

$$\begin{aligned}\int xe^x dx &= \int xd(e^x)\\ &= xe^x - \int e^x dx\\ &= xe^x - e^x.\end{aligned}$$

♣

Remark 1.1.1 Using this technique with Mathematical induction, one can evaluate

$$\int x^n e^x dx, \int x^n \sin x \; dx, \int p(x)e^x dx, \text{ etc.}$$

for any polynomial $p(x)$.

Example 1.1.2 Find (an explicit form of) $\displaystyle\int \ln x \; dx$.

Solution.

$$\begin{aligned}\int \ln x \; dx &= x \ln x - \int x \; d(\ln x) \\ &= x \ln x - \int dx \\ &= x \ln x - x + C.\end{aligned}$$

♣

Example 1.1.3 Find $\displaystyle\int e^{2x} \sin 3x \; dx$.

Solution.

$$\begin{aligned}\int e^{2x} \sin 3x \; dx &= \frac{1}{2} \int \sin 3x \; d(e^{2x}) \\ &= \frac{1}{2}[e^{2x} \sin 3x - \int e^{2x} d(\sin 3x)] \\ &= \frac{1}{2}e^{2x} \sin 3x - \frac{3}{2} \int (\cos 3x)e^{2x} \; dx \\ &= \frac{1}{2}e^{2x} \sin 3x - \frac{3}{4} \int (\cos 3x) d(e^{2x}) \\ &= \frac{1}{2}e^{2x} \sin 3x - \frac{3}{4}e^{2x} \cos 3x - \frac{9}{4} \int e^{2x} \sin 3x \; dx.\end{aligned}$$

Therefore,

$$(1 + \tfrac{9}{4}) \int e^{2x} \sin 3x \; dx = \frac{1}{2}e^{2x} \sin 3x - \frac{3}{4}e^{2x} \cos 3x + C$$

or equivalently,

$$\int e^{2x}\sin 3x\,dx = e^{2x}\Big(\frac{2}{13}\sin 3x - \frac{3}{13}\cos 3x\Big) + C.$$

♣

Example 1.1.4 Find $\int (3x-4)^3 x\,dx$.

Solution.

$$\begin{aligned}
\int (3x-4)^3 x\,dx &= \int (3x-4)^3 d(\tfrac{x^2}{2}) \\
&= (3x-4)^3\tfrac{x^2}{2} - \int 9(3x-4)^2\tfrac{x^2}{2}dx \\
&= (3x-4)^3\tfrac{x^2}{2} - \frac{3}{2}\int (3x-4)^2 d(x^3) \\
&= (3x-4)^3\tfrac{x^2}{2} - \frac{3}{2}(3x-4)^2x^3 - 3^2x^3(3x-4) \\
&\qquad + 3^2\int (3x-4)d(\tfrac{x^4}{4}) \\
&= (3x-4)^3\tfrac{x^2}{2} - \frac{3}{2}(3x-4)^2x^3 - 3^2x^3(3x-4) \\
&\quad + 3^2\int (3x^4-4x^3)dx
\end{aligned}$$

which is now easy to finish. (Of course, the whole thing could have been computed by expanding the integrand and integrate each monomial separately! Also, the solution above may be one of the worst. But I just wanted to show that it can be done by integration by parts.) ♣

1.2 Integration by substitution

The fundamental theorem is:

Theorem 1.2.1

$$\int f'(u)du = \int f'(u(x))u'(x)dx = f(u(x)) = f(u).$$

Proof. Immediate from

$$\frac{d}{dx}f(u(x)) = f'(u(x))u'(x).$$

□

1.2.1 Some direct results

Example 1.2.1 Verify that $\displaystyle\int \frac{dx}{x} = \ln|x| + C.$

The student may need to consider the cases $x < 0$ and $x > 0$ separately.

Example 1.2.2 Find $\displaystyle\int \frac{dx}{(x-a)^m}$, for every integer m.

Example 1.2.3 Find $\displaystyle\int \frac{x}{a^2+x^2}dx.$

Example 1.2.4 Find $\displaystyle\int \tan x \; dx.$

Exercises. Find an explicit form of the following indefinite integrals:

1. $\int \sec x \; dx$. [Hint: Consider $(\sec x)'$ and $(\tan x)'$. Then try to extract $\sec x$ from there. One should arrive at $\sec x = (\sec x + \tan x)'/(\sec x + \tan x)$.]

2. $\int \sec^2 x \; dx$

3. $\int \sec^3 x \; dx$

1.3 Further techniques

1.3.1 Substitution by trigonometric functions

Example 1.3.1 Find $\displaystyle\int \frac{dx}{3^2 + x^2}$.

Example 1.3.2 Find $\displaystyle\int \frac{dx}{\sqrt{1 - x^2}}$.

Example 1.3.3 Find $\displaystyle\int \frac{dx}{\sqrt{1 + x^2}}$.

Example 1.3.4 Find $\displaystyle\int \sqrt{\tan x}\, dx$.

Example 1.3.5 Find $\displaystyle\int \sin^m x \cos^n x \, dx$.

1.3.2 Partial fraction techniques

Fundamental theorem of Algebra implies:

Theorem 1.3.1 *Let $P(x)$ be a polynomial of degree $m \geq 1$ with real coefficients. Then P is the product of finitely many polynomials of degree not more than 2. Moreover, it can be arranged that the second degree polynomial factors do not admit any real roots.*

The following theorem is also well known.

Theorem 1.3.2 (Partial fractions) *Let $P(x), Q(x)$ be nonconstant polynomials with real coefficients. Assume that there are distinct real numbers $r_1, \ldots, r_k$ with positive integers $m_1, \ldots, m_k$ and the second degree polynomials $q_1, \ldots, q_\ell$ of type $x^2 + ax + b$ (with a, b real) with no real roots such that*

$$Q(x) = (x - r_1)^{m_1} \cdots (x - r_k)^{m_k} \cdot (q_1(x))^{k_1} \cdots (q_\ell(x))^{k_\ell}.$$

Then

$$\frac{P(x)}{Q(x)} = F_0(x) + \sum_{i=1}^{k} F_i(x) + \sum_{j=1}^{\ell} G_j(x),$$

where:

(1) $F_0(x)$ is a polynomial of degree equal to $\deg P - \deg Q$.

(2) For each $i = 1, \ldots, k$, $F_i(x)$ is of the form

$$\frac{a_{i1}}{(x - r_i)} + \frac{a_{i2}}{(x - r_i)^2} + \cdots + \frac{a_{im_i}}{(x - r_i)^{m_i}}$$

for some real constants $a_{i1}, \ldots, a_{im_i}$.

(3) For each $j = 1, \ldots, \ell$, $G_j(x)$ is of the form

$$\frac{p_{j1}(x)}{q_j(x)} + \frac{p_{j2}(x)}{(q_j(x))^2} + \cdots + \frac{p_{jk_j}(x)}{(q_j(x))^{k_j}},$$

where $q_j(x)$ a real quadratic polynomial with two distinct nonreal roots $\beta_j, \bar{\beta}_j$ and where each $p_{j\mu}(x)$ is a real polynomial given by

$$p_{j\mu}(x) = \alpha_{j\mu}(x - \bar{\beta}_j)^\mu + \bar{\alpha}_{j\mu}(x - \beta_j)^\mu$$

for some complex constant $\alpha_{j\mu}$.

We point out that for the purpose of integrating each term appearing in (3) above, one would like to use

$$\frac{p_{j\mu}(x)}{(q_j(x))^\mu} = \frac{\alpha_{j\mu}(x - \bar{\beta}_j)^\mu + \bar{\alpha}_{j\mu}(x - \beta_j)^\mu}{(x - \beta_j)^\mu (x - \bar{\beta}_j)^\mu} = \frac{\alpha_{j\mu}}{(x - \beta_j)^\mu} + \frac{\bar{\alpha}_{j\mu}}{(x - \bar{\beta}_j)^\mu}.$$

The student may integrate each term as if the complex constants were real numbers.[1]

Example 1.3.6 Find $\displaystyle\int \frac{2x - 2}{x^2 + 2x + 5}\,dx.$

Altogether we now have learned how to evaluate the indefinite integral of any rational functions in x variable.

[1]As one would guess at this point, the partial fraction theorem is a result of complex function theory (i.e., Complex Analysis). Complex function theory also guarantees the computational technique that these lecture notes relate.

1.3.3 Half-angle technique

If one uses the substitution

$$u = \tan \frac{x}{2},$$

then

$$\sin x = \frac{2u}{1+u^2}, \qquad \cos x = \frac{1-u^2}{1+u^2}, \qquad \tan x = \frac{2u}{1-u^2}.$$

Moreover,

$$dx = \frac{2\ du}{1+u^2}.$$

Now one can turn the indefinite integral

$$\int \frac{dx}{A + B \sin x}$$

into the integral of a rational function of $u = \tan \frac{x}{2}$ substitution.

Example 1.3.7 Find $\int \frac{3\ dx}{5-2\cos x}$.

1.3.4 Euler's method (Optional)

L. Euler considered evaluation problem for

$$A := \int R(x, \sqrt{ax^2 + bx + c})\ dx,$$

where R is a rational function.

We shall learn Euler's methods here in three cases:

Case I $a > 0$:
Put $u \pm \sqrt{a}x = \sqrt{ax^2 + bx + c})$. Then

$$x = \frac{c - u^2}{\pm 2u\sqrt{a} - b}$$

and hence $dx = r(u)du$, where $r(u)$ is a rational function of u. Either one of positive or negative sign can be chosen, and the indefinite integral A turns into an integral of a rational function, which in theory can be computed via the partial fraction technique. The method comes in three cases, not mutually exclusive.

Case II $c > 0$:

Put $ux \pm \sqrt{c} = \sqrt{ax^2 + bx + c}$). Then solve for x here, to obtain

$$x = \frac{\pm 2u\sqrt{c} - b}{a - u^2}.$$

Again a similar effect will take place, and either the positive or the negative sign can be taken.

Case III $ax^2 + bx + c = a(x - \alpha)(x - \beta)$, with both α and β real.

Put

$$(x - \alpha)u = \sqrt{a(x - \alpha)(x - \beta)}.$$

Then $x = \dfrac{a\beta - \alpha u^2}{a - u^2}$, and again similar effect takes place.

Example 1.3.8 Find $\displaystyle\int \frac{dx}{x\sqrt{x^2 + 4x - 4}}$.

Example 1.3.9 Find

$$\int \frac{dx}{x^2\sqrt{-x^2 + x + 2}}.$$

Example 1.3.10 Find

$$\int \frac{x^2\, dx}{\sqrt{-x^2 + 3x - 2}}.$$

Chapter 2

Functions by Power Series

2.1 Power series

The power series is introduced as the formal infinite sum

$$\begin{aligned} S(x) &= \sum_{m=0}^{\infty} a_m (x-c)^m \\ &= a_0 + a_1(x-c) + a_2(x-c)^2 + \ldots + a_k(x-c)^k + \ldots, \end{aligned}$$

where a_k's as well as c are real numbers. It has no particular meaning at this point, until one wants to understand it as a function of the variable x. In order to do that, one will consider replacing x by a specific number, say 3.

For instance, if our formal power series was given as

$$S(x) = \sum_{m=0}^{\infty} \frac{1}{m^3}(x-1)^m,$$

then one wants to know whether $S(3)$ makes sense, i. e., whether

$$S(3) = \sum_{m=0}^{\infty} \frac{1}{m^3} 2^m$$

defines a real value. If it does, or more generally if $S(r)$ defines a real value for some real values of r, then it makes sense to gather all such r's for which $S(r)$ yields a unique real value. If we denote by R the set of such values, then suddenly the power series $S(x)$ acquires significance and obtains the status of a real-valued function of the real variable x, which traverses over R.

Thus it is natural to learn how to find this set

$$R = R_S := \{r \in \mathbb{R} \colon S(r) \text{ defines a unique value in } \mathbb{R}\}.$$

And, to do that, one has to learn when any real numerical series

$$\sum_{m=0}^{\infty} A_m,$$

with A_m real, defines a unique real value.

2.2 Infinite numerical series

Definition 2.2.1

$$\sum_{m=0}^{\infty} A_m = \lim_{N\to\infty} \sum_{m=0}^{N} A_m.$$

To be more precise and logical, the meaning of the identity is as follows: If the right hand side exists as a finite value L, for instance, we say that $\sum_{m=0}^{\infty} A_m$ *converges to* L, and in such a case, we write

$$\sum_{m=0}^{\infty} A_m = L.$$

In case that the limit value of the right hand side diverges to $+\infty$ (or, to $-\infty$, respectively), we still use the expression

$$\sum_{m=0}^{\infty} A_m = \pm\infty,$$

respectively, following the conventional way.

Remark 2.2.1 (Limit) Calculus, the theory of functions and their differentiation/integration, is built upon the concept of *limit*, on top of four basic operations (addition, subtraction, multiplication and division). Due to the pedagogical reasons (not the mathematical reasons), discussion of the definition of limits and subsequent verifications and proofs are usually deferred to a later time, in the college mathematical education of "modern days". We are following a similar path, but we defer it to the last part of these notes. Interested readers may jump ahead to the early part of Chapter 5 to taste the precise definition and basic treatments of the limit.

2.3 Convergence of a series

The main goal of this section is to discuss the convergence (= the existence of the finite value limit) of a given series $\sum a_m$. As one may say, why do we not go directly about finding the exact value of the limit, rather than just trying to discuss whether the limit exists.

Finding the value is nice (especially because it says to a mathematician that the convergence has already been verified), but it does not happen very often. On the other hand, finding partial sums is "easier" as all that is required is just addition (including subtraction), which can be done by a machine in these days. Then, if one knows the convergence, then the results of these partial sums provide approximate values, arbitrarily close to the actual value. One of the main purposes of considering series is to find an approximate value, especially when the actual "true" value is difficult to obtain. Thus, deciding whether the series converges and, if it does, estimating how big the error may emerge as the important issue.

The exposition begins with the next subsection. But first, we verify the following important result.

Theorem 2.3.1 (n-th term divergence test) *If* $\lim_{n\to\infty} A_n \neq 0$, *then the series* $\sum A_n$ *diverges.*

Proof. Assume the contrary that there is the real number L satisfying $\lim_{m\to\infty} \sum_{k=0}^{N} A_k = L$. Then

$$A_n = \sum_{k=0}^{n} A_k - \sum_{k=0}^{n-1} A_k$$

and, consequently,

$$\begin{aligned}
\lim_{n\to\infty} A_n &= \lim_{n\to\infty} \Big(\sum_{k=0}^{n} A_k - \sum_{k=0}^{n-1} A_k \Big) \\
&= \lim_{n\to\infty} \sum_{k=0}^{n} A_k - \lim_{(n-1)\to\infty} \sum_{k=0}^{n-1} A_k \\
&= L - L = 0.
\end{aligned}$$

This contradicts the hypothesis. Thus the proof follows. $\square$

On the other hand, the converse of this theorem is not true, which we shall discuss a bit later.

2.3.1 Positive term series

Definition 2.3.1 $\sum A_m := \sum_{m=0}^{\infty} A_m$ is called a *positive-term series*, if $A_m \geq 0$, for every m.

Theorem 2.3.2 *If $\sum A_m$ is a positive term series and if there exists a constant $C > 0$ such that*

$$\sum_{m=0}^{N} A_m \leq C$$

for any positive integer N, then $\sum A_m$ converges to a value not exceeding C.

Note that the sequence of partial sums increase (in fact, nondecreasing, to say precisely) as the more positive (precisely speaking, nonnegative) terms are added. If the sequence of partial sums, an increasing sequence of real numbers, is bounded from above by a finite number, then it is intuitively obvious that the sequence must converge.

However, it turns out that this "obvious" understanding is difficult to prove mathematically! From the ancient days, the following were known.

Definition 2.3.2 Let A be a subset of the set $\mathbb{R}$ of real numbers. If there is a number $u \in \mathbb{R}$ such that

$$a \leq u, \quad \forall a \in A,$$

then u is called an *upper bound* of A.

If a set $A \subset \mathbb{R}$ has an upper bound, then A is said to be *bounded from above* (or simply, *bounded above*. Moreover, if a subset A of $\mathbb{R}$ is bounded above, the concept of the least upper bound is defined.

Definition 2.3.3 For $A \subset \mathbb{R}$ bounded above, an element $v \in \mathbb{R}$ is called the *least upper bound* of A if the following hold:

(1) $a \leq v$ for any $a \in A$. Namely, v is an upper bound.

(2) $v \leq u$ holds, whenever $a \leq u$ for any $a \in A$.

The least upper bound is usually called the *supremum* of A. The terminology makes sense, since the supremum is larger (precisely speaking, not smaller) than any element of A and since it is almost an element of A. The notation is usually as follows:

$$\text{lub } A = \sup A = \text{the least upper bound of } A.$$

For example, if $A = \{x \in \mathbb{R} \colon 0 < x < 1\}$, then the supremum of A is 1. If A were the closed interval $[0, 1]$ = the set of all $x \in \mathbb{R}$ with $0 \le x \le 1$ then the supremum of the set A is still 1. In the former case the supremum was not an element of A, but the latter case it was. In the latter case, the supremum is called the *maximum* of A. Likewise, the concepts of *bounded from below*, *lower bound*, and the *greatest lower bound* (or, the *infimum*) are defined and understood. Also the notation such as

$$\inf A, \text{glb } A, \max A, \min A$$

are commonly used.

Remark 2.3.1 The least upper bound is therefore the least of upper bounds. If the set A is empty, then any real number is an upper bound. Then what should be the least upper bound of $A = \varnothing$? According to the definition above there is no real number for that. Conventionally, mathematicians say, it is $-\infty$. Then the supremum of the empty set is $-\infty$, and the infimum of the empty set is $+\infty$. Such things do happen with the empty set!

We usually denote by $\sup A$ the supremum of A, and by $\inf A$ the infimum. Then, traditionally, the following axiom is accepted[1]:

> **Least Upper Bound Property:** *Any nonempty set $A \subset \mathbb{R}$, bounded above, has the least upper bound.*

Theorem 2.3.3 *Let $\varnothing \neq A \subset \mathbb{R}$. If A is bounded from above, then there is a unique least upper bound of A.*

[1] A more essential question may be "What is the real number?" If one describes the real number correctly, would the least upper bound property be possible to prove (and then it could become a theorem). This is the achievement of Dedekind, which one can learn from the famous book entitled *Principles of mathematical analysis*, by W. Rudin.

We leave the proof as an exercise to the reader, since it is just a logical deduction.

Well, to some students, the discussion up to this point was too formal. But if one wants to establish mathematics logically precisely, these have turned out to be necessary! Now accepting these, let us continue with

Proof of Theorem 2.3.2. It follows by the discussion above, and the definition of the limit. But, alas! We have never discussed the definition of limit, and so there is no satisfactory proof we can provide here. So again, we have to accept this theorem here, as such(!), although it is not so difficult to prove it. (This again shows the necessity of contents of Chapter 5 of this book.) □(!)

Here is a useful consequence.

Theorem 2.3.4 *For $r \geq 0$, the series $\sum r^n$ converges if, and only if, $r < 1$.*

Proof. Let $S_m = \sum_{n=1}^{m} r^n = 1 + r + r^2 + \ldots + r^m$. Then

$$S_m - rS_m = 1 - r^{m+1}.$$

Therefore, if $0 \leq r < 1$, $S_m = \frac{1-r^{m+1}}{1-r}$ and

$$\lim_{m \to \infty} S_m = \frac{1}{1-r}.$$

So the conclusion follows. □

2.4 Absolute convergence

Consider now the series $\sum B_n$ with $B_n \in \mathbb{R}$. But we allow the general term B_n can take positive or negative values.

Definition 2.4.1 (Absolute convergence) The series $\sum B_n$ is said to *converge absolutely*, if the absolute value term series $\sum |B_n|$ converges.

This definition shows some usefulness due to the following

Theorem 2.4.1 *The series $\sum B_n$ converges if $\sum |B_n|$ converges. In word, any absolutely convergent series converges.*

The reason why this theorem is true can be seen from (again this is not a precise proof, logically speaking):

$$\begin{aligned} &\sum_{n=0}^{\infty} B_n \text{ converges} \\ &\quad \Leftrightarrow \lim_{m\to\infty}\Big(\sum_{n=0}^{\infty} B_n - \sum_{n=0}^{m} B_n\Big) = 0 \\ &\quad \overset{!}{\Leftrightarrow} \quad \lim_{m\to\infty} \sum_{n=m+1}^{\infty} B_n = 0 \\ &\quad \Leftrightarrow \lim_{m\to\infty} \Big|\sum_{n=m+1}^{\infty} B_n\Big| = 0. \end{aligned}$$

If $\sum |B_n|$ converges, then in particular, $\lim_{m\to\infty} \sum_{n=m+1}^{\infty} |B_n| = 0$. By triangle inequality and the monotonicity of limit, one arrives at

$$\lim_{m\to\infty} \Big|\sum_{n=m+1}^{\infty} B_n\Big| \le \lim_{m\to\infty} \sum_{n=m+1}^{\infty} |B_n| = 0.$$

Thus the claim follows. [2] □

[2]This argument, as it is written here, has a gap, although it has been completely filled. The second $\Leftrightarrow$ still needs proof, (What exactly is the problem here? Think about it!) for $\Leftarrow$. One needs to check that any *Cauchy sequence* of real numbers has to converge (which is known to be true).

Proof of Theorem above by Least upper bound property: Lang's Short Calculus (our textbook) gives a clever proof and so I would like to present that here.

For every n, let

$$B_n^+ = \begin{cases} B_n & \text{if } B_n \geq 0 \\ 0 & \text{if } B_n < 0, \end{cases}$$

and

$$B_n^- = -(B_n - B_n^+).$$

Then $B_n = B_n^+ - B_n^-$ and $B_n^+ \geq 0$ and $B_n^- \geq 0$ for every n. Moreover,

$$\sum_{n=0}^{m} B_n^+ \leq \sum_{n=0}^{m} |B_n| \quad \text{and} \quad \sum_{n=0}^{m} B_n^- \leq \sum_{n=0}^{m} |B_n|.$$

By the hypothesis, $\sum |B_n|$ converges, and hence both $\sum B_n^+$ and $\sum B_n^-$ converge, respectively to B^+ and B^-, say. Then

$$\begin{aligned} \left| \sum_{n=0}^{m} B_n - (B^+ - B^-) \right| &= \left| \sum_{n=0}^{m} (B_n^+ - B_n^-) - (B^+ - B^-) \right| \\ &= \left| \left(\sum_{n=0}^{m} B_n^+ - B^+ \right) - \left(\sum_{n=0}^{m} B_n^- - B^- \right) \right| \\ &\leq \left| \sum_{n=0}^{m} B_n^+ - B^+ \right| + \left| \sum_{n=0}^{m} B_n^- - B^- \right|. \end{aligned}$$

And, we have seen that this last converges to 0. Therefore, $\sum B_n$ converges and

$$\sum_{n=0}^{\infty} B_n = B^+ - B^-.$$

□

Remark 2.4.1 There are series $\sum C_n$ of real numbers that are convergent, while it fails to converge absolutely. $\sum (-1)^n n^{-1}$ is an example. Of course one needs verification. We shall analyze this types of series later. On the other hand, these series has an interesting property that the limit values as well as convergence can be altered when the order of summation is rearranged. [Finitely many changes will not alter anything about convergence or the limit values, however.]

Convergent series that are not absolutely convergent are said to be *conditionally convergent.*

2.5 Tests for Convergence

The main directive of this section is to develop several effective criteria for the convergence of infinite series, based upon the discussions from the preceding sections.

2.5.1 Comparison test

Theorem 2.5.1 (Comparison test) *Let $\sum a_n$ and $\sum b_n$ be two positive term series. If there exists a constant $C > 0$ such that $a_n \leq Cb_n$ for every $n = 0, 1, \ldots$, and if $\sum b_n$ converges, then $\sum a_n$ converges also.*

Proof. Let m be a positive integer. Let $S_m = \sum_{n=0}^{m} a_m$ and $T_m = \sum_{n=0}^{m} b_m$. Then

$$S_m \leq \sum_{n=0}^{m} Cb_m \leq C \sum_{n=0}^{\infty} b_m.$$

Since $\sum b_n$ converges, let $T = \sum b_n = \lim_{m\to\infty} T_m$, which is a finite positive number. Then the above implies that

$$S_m \leq CT, \quad \forall m = 0, 1, \ldots.$$

Now, $\lim_{m\to\infty} S_m = \sum a_n$ converges by Theorem 2.3.2. □

Example 2.5.1 Show that $\sum_{n=1}^{\infty} \frac{2^n}{5^{n-2}-2}$ converges.

Solution. Note that

$$\begin{aligned}\sum_{n=1}^{\infty} \frac{2^n}{5^{n-2}-2} &= -\frac{10}{9} - 2 + \sum_{n=3}^{\infty} \frac{2^n}{5^{n-2}-2} \\ &= -\frac{28}{9} + 4\sum_{n=3}^{\infty} \frac{2^{n-2}}{5^{n-2}-2} \\ &= -\frac{28}{9} + 4\sum_{n=1}^{\infty} \frac{2^n}{5^n-2}.\end{aligned}$$

Since $5^n - 2 > 4^n$ for any positive integer n, we have

$$\frac{2^n}{5^n-2} \leq \left(\frac{1}{2}\right)^n.$$

Also $\sum \left(\frac{1}{2}\right)^n$ converges (by Theorem 2.3.4). Altogether, we see that $\displaystyle\sum_{n=1}^{\infty} \frac{2^n}{5^{n-2}-2}$ converges. $\square$

Remark 2.5.1 Needless to say that if two positive term series $\sum a_n$ and $\sum b_n$ admit a constant $C > 0$ with $a_n \leq Cb_n$ for all $n > 1$, and if $\sum a_n$ diverges (i. e., $\sum a_n = \infty$), then $\sum b_n$ also diverges. This obvious fact is also included in the Comparison Test.

2.5.2 Limit comparison test

Theorem 2.5.2 (Limit comparison test) *If two positive term series $\sum a_n$ and $\sum b_n$ admit two positive constants c, C such that*

$$0 < c \leq \lim_{n\to\infty} \frac{a_n}{b_n} \leq C,$$

then either both $\sum a_n$ and $\sum b_n$ converge, or both of them diverge.

Proof. Let $\lim_{n\to\infty} \frac{a_n}{b_n} = L$. The hypothesis of Theorem says $c \leq L \leq C$, Then a_n/b_n is very close to T, whenever n is sufficiently large. This means that there exists some positive integer n_0 such that $c/2 < a_n/b_n < 2C$ whenever $n > n_0$. [3] Since $c/2, 2C$ are finite positive numbers, the claim follows by the comparison test in the preceding subsection. $\square$

Example 2.5.2 Show that $\sum_{n=1}^{\infty} \frac{2^n}{5^{n-2}-2}$ converges.

Solution. Let $n \geq 3$. Note that $\sum_{n=3}^{\infty} \frac{2^n}{5^{n-2}-2}$ is a positive term series. Also if we let

$$a_n = \frac{2^n}{5^{n-2}-2}, \quad b_n = \frac{2^n}{5^{n-2}} = 4(2/5)^{n-2}.$$

Then $\lim_{n\to\infty} a_n/b_n = 1$. Since $\sum b_n$ converges, we conclude that $\sum a_n$ converges as well. $\square$

Note that this may simplify some steps from the comparison test.

[3]This appeals to the intuitive understanding of the limit, for now. It can be made precise, using the definition of the limit. So the reader should realize why one would like to understand the accurate meaning of limits and how to use it.

2.5.3 Ratio test

Theorem 2.5.3 (The Ratio Test I) *Let $\sum a_n$ be a positive term series. Then,*

$$\lim_{n\to\infty} \frac{a_{n+1}}{a_n} = r < 1$$

then $\sum a_n$ converges.

Proof. This follows from the definition of the limit and the comparison test. Since the terms $\frac{a_{n+1}}{a_n}$ are arbitrarily close to r, for any small number $\epsilon > 0$ there exists a positive integer N such that the value of $\frac{a_{n+1}}{a_n}$ is between $r - \epsilon$ and $r + \epsilon$, whenever $n > N$. This is the meaning of the casual statement that, as n becomes arbitrarily large, $\frac{a_{n+1}}{a_n}$ are arbitrarily close to r. We take a special case $\epsilon = \frac{1-r}{2}$. Then,

$$\frac{a_{n+1}}{a_n} - r < \frac{1-r}{2},$$

namely,

$$\frac{a_{n+1}}{a_n} < \frac{1+r}{2}$$

whenever $n \geq N + 1$. If we let $s = (1 + r)/2$, then this means

$$a_{n+k} < a_{n+k-1}s < a_{n+k-2}s^2 < \cdots < a_n s^k, \ \forall k = 1, 2, \ldots.$$

So

$$\sum a_n < \sum_{n=0}^{N} a_n + a_{N+1} \sum_{k=0}^{\infty} s^k.$$

Notice that $0 < s < 1$ since $0 < r < 1$. Therefore

$$\sum a_n < \sum_{n=0}^{N} a_n + \frac{a_{N+1}}{1-s}.$$

Since the right hand side is bounded, the comparison test implies that $\sum a_n$ converges. □

Theorem 2.5.4 (The Ratio Test II) *For the positive series $\sum a_n$, if*

$$\lim_{n\to\infty} \frac{a_{n+1}}{a_n} = r > 1$$

then $\sum a_n$ diverges.

Sketch of the Proof. Again, the interpretation of the concept of limit in the hypothesis implies that there is a natural number $N > 0$ such that, whenever $n > N$,

$$\frac{a_{n+1}}{a_n} - r > \frac{1-r}{2}.$$

Put $s = \frac{1+r}{2}$, then $s > 1$ (since $r > 1$) and

$$a_{n+1} > a_n s, \ \forall n \geq N+1.$$

This implies, due to a similar (but different) reasoning as in the preceding proof argument,

$$\sum a_n > \sum_{n=0}^{N} a_n + a_{N+1} \sum_{k=0}^{\infty} s^k.$$

But, since $s > 1$ here, the right hand side is $+\infty$, and hence the desired conclusion follows. □

We leave checking the detail to the reader as an exercise.

Remark 2.5.2 In the two theorems above (The Ratio test I and II), the case when $\lim_{n\to\infty} \frac{a_{n+1}}{a_n} = 1$ was never discussed. In fact, the definition of this limit condition cannot give any further conclusion. Typical examples are the "harmonic series" $\sum \frac{1}{n}$ as well as "p-series" $\sum \frac{1}{n^p}$ with $p > 0$. The ratio test leaves their convergence inconclusive. These are however understood, and we shall discuss these a bit later.

Remark 2.5.3 Even if the series $\sum a_n$ were not a positive term series, $\sum |a_n|$ is. So one may apply the ratio test; if

$$\lim_{n\to\infty} \frac{|a_{n+1}|}{|a_n|} < 1$$

then the series $\sum a_n$ converges absolutely, and hence it converges. If the limit value is strictly larger than 1, the series diverges. But the reason is not by the ratio test, but by the n-th term divergence test. We leave the verification of this last statement to the reader.

The ratio test is extremely handy. Let us revisit the example above.

Example 2.5.3 Show that $\sum_{n=1}^{\infty} \frac{2^n}{5^{n-2}-2}$ converges.

Solution by the ratio test. The reader should see that

$$r := \lim_{n\to\infty} \frac{2^{n+1}}{5^{n-1}-2} \frac{5^{n-2}-2}{2^n} = \frac{2}{5} < 1$$

immediately. So the given series converges by the ratio test. ♣

2.5.4 The Root test

Theorem 2.5.5 (The Root Test) *Let $\sum a_n$ be a positive term series. If*

$$\lim_{n\to\infty} \sqrt[n]{a_n} = r < 1,$$

then $\sum a_n$ converges.

Proof. Assume that $r < 1$. Then again, the meaning of limit says that the value of $\sqrt[n]{a_n}$ approaches arbitrarily close to r. Hence if we let $r_1 := r/2$ and $r_2 := r + \frac{1-r}{2}$, then $r_1 < r < r_2$ and, there exists some sufficiently large integer n_0 such that

$$r_1 < \sqrt[n]{a_n} < r_2$$

whenever $n \geq n_0$. This inequality implies that

$$a_n < (r_2)^n$$

and the geometric series $\sum_{n=n_0}^{\infty}(r_2)^n$ converges. This implies that $\sum_{n=n_0}^{\infty} a_n$ converges and consequently, $\sum a_n = S_{n_0-1} + \sum_{n=n_0}^{\infty} a_n$ converges, as the partial sum $S_{n_0-1} := \sum_{n=0}^{n_0-1} a_n$ is finite. □

We also have

Theorem 2.5.6 *Let $\sum a_n$ be a positive term series. If*

$$\lim_{n\to\infty} \sqrt[n]{a_n} = r > 1,$$

then $\sum a_n$ diverges.

Proof. This is again by the n-th term divergence test. Further detail is left to be checked by the reader. □

Remark 2.5.4 As the convergence test, students usually find the ratio test as more versatile and easier than the root test. But the root test is worth while studying. In fact, when one applies these criteria in the context of power series, the root test has a definite merit. But that is subtle, and will not be emphasized on the level of this course.

2.5.5 Integral test and harmonic series

The Riemann integral is the definite integral essentially defined for the piece-wise continuous functions that are bounded, and the integral is built upon a finite interval. But then one may consider the extended concept such as

$$\int_0^\infty f(x)dx := \lim_{B\to\infty} \int_0^B f(x)dx.$$

Such combinations of integrals with limits are conventionally called the *improper integrals.* We shall not discuss too much about it, but we shall just use the example of type introduced above.

For instance

$$\int_1^\infty \frac{1}{x}dx = +\infty.$$

and

$$\int_1^\infty \frac{1}{x^2}\,dx = \lim_{B\to\infty}\Big[-\frac{1}{x}\Big]_1^B = 1$$

which is finite. Now, this implies a useful result.

Since $f(x) = \frac{1}{x}$ is a *decreasing, positive valued* function on $[1,\infty) := \{x \in \mathbb{R}: x \geq 1\}$, one sees that

$$f(2) + f(3) + \ldots \leq \int_1^\infty f(x)dx \leq f(1) + f(2) + \ldots.$$

The left hand side is the sum of the area of the rectangles built upon the interval $[n, n+1]$ with height $f(n+1)$ for every $n = 1, 2, \ldots$. Since every such rectangle is below the graph of $f(x)$, the first inequality holds. The second inequality follows, on the other hand, because the rectangles build upon the intervals $[n, n+1]$ with height $f(n)$, for $n = 1, 2, \ldots$.

Then, notice that the second inequality now implies

$$+\infty = \int_1^\infty f(x)dx \leq f(1) + f(2) + \ldots = \sum_{n=1}^\infty \frac{1}{n}.$$

So $\sum \frac{1}{n}$ diverges! Notice that this was the problem left inconclusive by the root test as well as the ratio test.

This generalizes immediately to a test (called the *integral test*) as follows:

Theorem 2.5.7 (Integral test) *If $a_n = f(n)$ for every $n = 1, 2, \dots$ for a function $f(x)$ defined on the line $[1, \infty) = \{x : x \geq 1\}$ satisfying*

- *$f(x) \geq 0$ for any $x \geq 1$,*
- *$f(x) \geq f(y)$ whenever $x < y$, i. e., f is decreasing,*

then $\sum a_n$ converges if, and only if, $\int_1^\infty f(x)dx$ converges.

Corollary 2.5.8 (Convergence of p-series) *For any real number p, the series $\sum \frac{1}{n^p}$ converges if and only if $p > 1$.*

For the proof of corollary, perhaps we need to discuss the case $p \leq 0$. In such cases, the general term $\frac{1}{n^p} = n^q$ if we put $q = -p \geq 0$. So the general term is not less than 1, for any n. Hence the series diverges by the n-th term divergence test. Other cases follow, of course, by the integral test.

2.5.6 Alternating series

Up to now, most discussions pertaining to convergence of series have been concentrated upon the positive term series. For the general series, typical recommendation is to use absolute convergence (which is stronger than just convergence) and the n-th term divergence test. Many series cannot be tested by these methods. Then we have one more useful criterion for convergence:

Theorem 2.5.9 (Alternating series criterion) *If $a_n \geq 0$ for every n satisfies*

(1) $a_{n+1} \leq a_n$ for every $n = 0, 1, \dots,$ and

(2) $\lim_{n\to\infty} a_n = 0$,

then the series $S = \sum_{n=0}^{\infty} (-1)^n a_n$ converges.

Proof. We use the method used earlier (without a precise reasoning) that the convergence of S follows if

$$\lim_{N\to\infty} \sum_{n=N}^{\infty} (-1)^n a_n = 0.$$

This last is easy to see by the hypothesis. If N is even, then

$$\sum_{n=N}^{\infty}(-1)^n a_n = \sum_{n=N}^{\infty}(a_n - a_{n+1}) \geq 0.$$

Therefore,

$$\begin{aligned}\Big|\sum_{n=N}^{\infty}(-1)^n a_n\Big| &= \sum_{n=N}^{\infty}(-1)^n a_n \\ &= a_N - \sum_{n=N+1}^{\infty}(a_n - a_{n+1}) \\ &\leq a_N.\end{aligned}$$

On the other hand, if N is odd,

$$\sum_{n=N}^{\infty}(-1)^n a_n = -\sum_{n=N}^{\infty}(a_n - a_{n+1}) \leq 0.$$

Hence,

$$\begin{aligned}\Big|\sum_{n=N}^{\infty}(-1)^n a_n\Big| &= -\sum_{n=N}^{\infty}(-1)^n a_n \\ &= a_N - \sum_{n=N+1}^{\infty}(a_n - a_{n+1}) \\ &\leq a_N.\end{aligned}$$

Both cases verify that $\lim_{N\to\infty}\sum_{n=N}^{\infty}(-1)^n a_n = \lim_{N\to\infty} a_N = 0$, yielding the desired conclusion. □

Example 2.5.4 Verify that the following series converge:

(1) $1 - \frac{1}{2} + \frac{1}{3} - \frac{1}{4} + \cdots + \frac{(-1)^n}{n} + \cdots$.

(2) $\sum(-1)^n \frac{2^{2n+1}}{(2n+1)!}$.

(3) $\sum \frac{(-1)^n}{n^3}$.

2.6 Radius of convergence of power series

Now we are ready to discuss how to solve the problem which we posed in the beginning of this chapter:

> *For the power series $\sum a_n(x-c)^n$, determine the set of values of x for which the power series converges to (and hence assigns) a unique value.*

This will allow us to consider (many) functions given in the form of infinite power series.

2.6.1 By the ratio test

If $x \in \mathbb{R}$ would satisfy the condition

$$\lim_{n\to\infty} \frac{|a_{n+1}(x-c)^{n+1}|}{|a_n(x-c)^n|} < 1, \tag{2.1}$$

then for such a value of x the series $\sum a_n(x-c)^n$ converges absolutely. On the other hand, if the limit value is > 1, then the series would diverge.

Thus we are motivated enough to analyze (2.1). We immediately see that

$$\begin{aligned} 1 &> \lim_{n\to\infty} \frac{|a_{n+1}(x-c)^{n+1}|}{|a_n(x-c)^n|} \\ &= \lim_{n\to\infty} \frac{|a_{n+1}|}{|a_n|} \cdot |x-c|. \end{aligned}$$

Let

$$R = \lim_{n\to\infty} \frac{|a_n|}{|a_{n+1}|} \tag{2.2}$$

then the value of x satisfies (2.1) if and only if

$$|x-c| < R.$$

On the other hand, if $|x-c| > R$, then it holds that

$$\lim_{n\to\infty} \frac{|a_{n+1}(x-c)^{n+1}|}{|a_n(x-c)^n|} > 1$$

and consequently the series $\sum a_n(x-c)^n$ diverges.

Summarizing the discussion so far, we arrive at

Theorem 2.6.1 (Radius of convergence by the ratio test) *For the power series* $S := \sum a_n(x-c)^n$ *with center* c*, let*

$$A := \{x\colon |x-c| < R\} \text{ and } B := \{x\colon |x-c| > R\},$$

where

$$R := \lim_{n\to\infty} \frac{|a_n|}{|a_{n+1}|}. \tag{2.3}$$

Then the series S *converges for any* $x \in A$ *and diverges for any* $x \in B$.

The open interval A coincides with the interval $(-R+c, c+R)$ of the radius R centered at c. On the other hand $B = \mathbb{R}\setminus[-R+c, c+R]$. Thus the ambiguity is at two endpoints of A; the convergence at those points should be discussed and decided (if possible) separately.

Definition 2.6.1 The set A is called the *open interval of convergence.* The power series S defines a function $S\colon A \to \mathbb{R}$. The center c of this interval A is called the *center of power series* S. Finally, the number R defined in the theorem above is called (rightly) the *radius of convergence* of the power series S.

2.6.2 By the root test

The exposition here shares the same principle as in the preceding section. $\sum a_n(x-c)^n$ converges if

$$\lim_{n\to\infty} \sqrt[n]{|a_n||x-c|^n} < 1, \tag{2.4}$$

and diverges if

$$\lim_{n\to\infty} \sqrt[n]{|a_n||x-c|^n} > 1, \tag{2.5}$$

Hence we analyze the inequality (2.4). We immediately see that this condition on x is the same as

$$|x-c| < R' := \left[\lim_{n\to\infty} \sqrt[n]{|a_n|}\right]^{-1}. \tag{2.6}$$

Of course, the condition on x given by (2.5) is equivalent to $|x-c| > R'$.

Summarizing, we have

Theorem 2.6.2 (Radius of convergence by the root test) *The power series $S = \sum a_n(x-c)^n$ converges at x if $|x-c| < R'$, and in contrast diverges at x whenever $|x-c| > R'$, where*

$$R' = \frac{1}{\lim_{n\to\infty} \sqrt[n]{|a_n|}}. \tag{2.7}$$

Remark 2.6.1 Again, the set $A' := \{x\colon |x-c| < R'\}$ is called the open interval of convergence, and c the center of the power series S. Of course, if both R and R' (defined in (2.3) and (2.7), respectively) exist, then $R = R'$.

[4]

2.6.3 Interval of convergence

Let us take a specific example. Consider the power series

$$S = \sum_{n=1}^{\infty} \frac{1}{n} x^n.$$

Then the center is at 0. The radius of convergence is given by

$$R = \lim_{n\to\infty} \frac{|n+1|}{|n|} = 1.$$

So the open interval of convergence is $A = \{x\colon -1 < x < 1\}$. As pointed out, we are only to check the convergence of S at $x = \pm 1$, since S diverges for any x with $|x| > 1$.

Then if $x = 1$, then

$$S|_{x=1} = \sum_{n=1}^{\infty} \frac{1}{n} = \infty$$

[4]In general the limits defining the radii of convergence need not exist. In such a case, there is a concept called superior and inferior limits (denoted by $\limsup_{n\to\infty} A_n$ and $\liminf_{n\to\infty} A_n$). These always exist (could be $\pm\infty$, too) and defines the radius of convergence. In such a case, the radius of convergence obtained by the root test is known to be the optimal (therefore, correct) one. But we shall not discuss this any more, on this level.

as we saw by the integral test. On the other hand, if $x = -1$,

$$S|_{x=-1} = \sum_{n=1}^{\infty} \frac{(-1)^n}{n}$$

and it converges, by the alternating series criterion. Therefore the maximum set of convergence for the power series S is the interval $[-1, 1) := \{x\colon\ -1 \leq x < 1\}$. This set is called the *interval of convergence* for S, and S defines a real-valued function on the interval of convergence $[-1, 1)$.

2.7 Termwise calculus

For the purpose of Calculus, i. e., theory of functions and their differentiation and integration, the fact that the power series $S = \sum a_n(x-c)^n$ being defined on the interval I, say, of convergence defines a function $f = S\colon I \to \mathbb{R}$ is not a satisfactory conclusion. We wish to investigate how smooth the function $f(x) = \sum a_n(x-c)^n$.

Since our theory of differentiable functions require the domain of definition be an open set, we naturally restrict ourselves to the function $f(x) = \sum a_n(x-c)^n$ defined on the open interval of convergence with positive radius, say R. Of course each partial sum $f_m(x) := \sum_{n=1}^{m} a_n(x-c)^n$ is a polynomial in the variable x, it is indefinitely many times differentiable, and has the primitive (= antiderivative = indefinite integral) inductively, also, indefinitely many times. However, $f(x) = \lim_{m\to\infty} f_m(x)$ may not be able to enjoy such privileges. As we already saw, even the domain of definition of f is no longer the whole real line (even though each f_m was well-defined on the whole real line, for every m) but just the (open) interval of convergence.

2.7.1 Termwise antidervative

Of course, the indefinite integral of a partial sum is obvious:

$$\int f_m(x)dx = \sum_{n=0}^{m} \frac{a_n}{n+1}(x-c)^{n+1}$$

for every positive integer m. But would it be true that

$$\int f(x)dx = C + \sum_{n=0}^{\infty} \frac{a_n}{n+1}(x-c)^{n+1}\ ?$$

It may be. Actually, it is. But we need a verification/proof, do we not? [5]

The following result can be checked immediately.

[5]Many people, as far as the author is aware of, feel and say openly that they do not care to learn the mathematical proofs, and they just want to know whether it is true. But that does not make any sense. What if they worked merrily with all such "facts" approved by a "trusted ones" and turned out much later that those "facts" were false in fact, and so your work based upon that had turned out useless? Proofs may be difficult to understand, but is necessary!

Proposition 2.7.1 *The radius of convergence of* $T = \sum_{n=0}^{\infty} \frac{a_n}{n+1}(x-c)^{n+1}$ *coincides with the radius of convergence, say* R*, of* $S = \sum_{n=0}^{\infty} a_n(x-c)^n$*. Thus both functions are well-defined on the interval* $A := (-R+c, c+R)$.

We omit the proof since it follows directly by the definition of the radius of convergence.

Moreover, we checked that the derivative of the partial sums of T coincides with the partial sums of S. This means that

$$\lim_{m\to\infty} T_m(x) = \lim_{m\to\infty} \int S_{m-1}(x)dx$$

but we still have to prove that

$$\lim_{m\to\infty} \int S_{m-1}(x)dx = \int \lim_{m\to\infty} S_{m-1}(x)dx. \tag{2.8}$$

Since we just checked in the proposition above that the last equals

$$\int S(x)dx$$

for any $x \in A$, it suffices to show that (2.8) holds.

Let $t \in A = (-R+c, c+R)$ with $t \neq c$. Then consider the closed interval

$$J_t = \{x \in \mathbb{R} \colon c \leq x \leq t \text{ or } t \leq x \leq c\}$$

with endpoints at c and t. Then by the Fundamental Theorem of Calculus, it holds for every continuous function $g \colon A \to \mathbb{R}$, that

$$\frac{d}{dx}\int_c^x g(s)ds = g(x), \quad \forall x \in J_t,$$

or equivalently the definite integral $\int_c^x f(s)ds$ is an antiderivative [6] of $f(x)$ at every $x \in A$.

---

[6]This is an important point! In the current highschool mathematics education, most textbooks of Korea do not teach the definition of definite integral at all, or worse, do it incorrectly! The definition of definite integral should be done by the way of understanding the *area/volume by slices* as Archimedes of Syracuse (ancient Sicily) discovered more than a couple of millenia ago. We shall discuss this in Chapter 5 of this book, briefly.

Now (2.8) can be understood via the definite integral. Let $R-|t-c|=\epsilon$. Then it is a positive number. Notice that $\lim_{m\to\infty}|a_m|/|a_{m+1}|=R$. Thus we may take a positive integer N such that $|a_{m+1}|<|a_m|/(R-\epsilon/2)$ whenever $m>N$. Thus, any integer $k>1$, we have

$$|a_{m+k}|<|a_m|(R-\frac{\epsilon}{2})^{-k}$$

and consequently, with $x\in J_t$

$$\begin{aligned}\sum_{n=m}^{\infty}|a_n||x-c|^n &\le \sum_{k=0}^{\infty}|a_{m+k}||t-c|^{m+k}\\ &\le |a_m||t-c|^m\sum_{k=0}^{\infty}\left(\frac{R-\epsilon}{R-\frac{\epsilon}{2}}\right)^k\\ &= \frac{2R-\epsilon}{\epsilon}|a_m||t-c|^m\end{aligned}$$

Since the last expression (which converges to 0) is independent of of the location of x in J_t, we may represent the last term as $\delta_m(t)$.

Now, for any $x\in J_t$, we have

$$\begin{aligned}&\Big|\int_c^x S(s)ds-\int_c^x S_{m-1}(s)ds\Big|\\ &\qquad=\Big|\int_c^x\sum_{n=m}^{\infty}|a_m||s-c|^m ds\Big|\\ &\qquad\le\Big|\int_c^t\sum_{n=m}^{\infty}|a_m||s-c|^m ds\Big|\\ &\qquad\le\Big|\int_c^t\delta_m(t)ds\Big|\\ &\qquad=\delta_m(t)\cdot|t-c|\\ &\qquad\to 0,\end{aligned}$$

as $m\to\infty$. This proves (2.8). □

As a direct result, we obtain:

Theorem 2.7.1 (Termwise antiderivative for power series) *If the power series* $S = \sum_{n=0}^{\infty} a_n x^n$ *has positive radius of convergence* R*, then the series* $T = \sum_{n=0}^{\infty} \frac{a_n}{n+1} x^{n+1}$ *has the same radius of convergence and the identity*

$$\frac{d}{dx} T(x) = S(x)$$

holds whenever the variable x *takes the value in* $A = (-R+c, c+R)$*. Equivalently, under the same restriction on* x *and radii of convergence*

$$\int \left(\sum a_n x^n \right) dx = \sum \int a_n x^n dx.$$

2.7.2 Termwise derivative

The termwise antiderivative has been shown to coincide with the actual antiderivative, in the preceding subsection. What about the derivative?

Theorem 2.7.2 (Termwise differentiation) *If the power series* $S = \sum_{n=0}^{\infty} a_n (x-c)^n$ *has positive radius of convergence* R*, then the series* $T = \sum_{n=0}^{\infty} n a_n (x-c)^{n-1}$ *has the same radius of convergence, and satisfies*

$$\frac{d}{dx} \sum_{n=0}^{\infty} a_n (x-c)^n = \sum_{n=0}^{\infty} \frac{d}{dx} a_n (x-c)^n = \sum_{n=0}^{\infty} n a_n (x-c)^{n-1}.$$

The proof uses the preceding theorem: Start with

$$\frac{d}{dx} \sum \frac{a_n}{n+1} (x-c)^{n+1} = \sum a_n x^n.$$

Then let $b_{n+1} = \frac{a_n}{n+1}$ and $m = n+1$, then it follows immediately that

$$\frac{d}{dx} \sum b_m (x-c)^m = \sum m b_m (x-c)^{m-1},$$

which yields the desired conclusion. □

Chapter 3

Taylor development

3.1 Mean-value theorem and Taylor's theorem

Recall

Theorem 3.1.1 (Mean-Value Theorem) *Let $a < b$. If $f\colon [a,b] \to \mathbb{R}$ is continuous, and if the restriction $f|_{(a,b)}\colon (a,b) \to \mathbb{R}$ is continuously differentiable, then there exists a point $c \in (a,b)$ such that*

$$\frac{f(b) - f(a)}{b - a} = f'(c).$$

Let $r > 0$ and let $b \in (-r+a, a+r)$. Notice that for any $x \in (a,b) \cup (b,a)$, the Mean-Value Theorem says that there exists $c_1 \in (a,x) \cup (x,a)$ such that

$$\frac{f(x) - f(a)}{x - a} = f'(c_1).$$

This implies that

$$f(x) = f(a) + f'(c_1)(x - a). \tag{3.1}$$

On the other hand, it seems to show that $f(x)$ is a linear function! But this cannot be true since there are infinitely many functions that are not linear but continuously differentiable. Then what does (3.1) say? Notice first that c_1 depends upon x. The relation between c_1 and x should be as complicated as the (nonlinear) function $f(x)$. Nevertheless the second point is worth mentioning: the function f can be efficiently approximated by a linear function in a sufficiently small interval, which we knew from the definition of differentiability of f.

Then we may extend the discussion, if f is of class C^2, namely, f is twice differentiable at every point of $(-r+a, a+r)$ for some $r > 0$ and the second derivative $f''(x)$ is continuous there. Then let

$$R_2(x) := f(x) - [f(a) + f'(a)(x-a)], \tag{3.2}$$

and try to find $c_2 \in (x, a) \cup (a, x)$, depending upon x, which satisfies

$$R_2(x) = \frac{f''(c_2)}{2!}(x-a)^2.$$

This is a special case of a theorem by Lagrange[1] stated below.

Definition 3.1.1 For a function $f(x)$ of class C^k defined on an open interval $(a-r, a+r) := \{x \in \mathbb{R} \colon |x-a| < r\}$ for some $r > 0$, the function

$$P_m(x) = f(a) + f'(a)(x-a) + \frac{f''(a)}{2!}(x-a)^2 + \ldots + \frac{f^{(m)}(a)}{m!}(x-a)^m,$$

for $m = 0, 1, \ldots, k$, is called the *m-th Taylor polynomial* of f at a. Moreover, the function

$$R_m(x) := f(x) - P_m(x)$$

is called the *m-th remainder* of f at a.

Theorem 3.1.2 (Lagrange) *If $f \in C^{k+1}(a-r, a+r)$, then for any $x \in (a-r, a+r)$ there exists $c_m \in (x, a] \cup [a, x)$ satisfying*

$$R_m(x) = \frac{f^{(m+1)}(c_m)}{(m+1)!}(x-a)^{m+1}$$

for any $m \in \{0, 1, \ldots, k\}$.

We will present a proof shortly. But, we present now the following:

[1] Joseph-Louis Lagrange (1736-1813), France.

Theorem 3.1.3 (Taylor expansion by Lagrange) *Let $a < b$ and let $f\colon (a,b) \to \mathbb{R}$ is of class C^k, for some positive integer k. Then for any $p, x \in (a,b)$, there exists $q \in (x,p] \cup [p,x)$ such that*

$$f(x) = \sum_{n=1}^{k-1} \frac{f^{(n)}(p)}{n!}(x-p)^n + \frac{f^{(k)}(q)}{k!}(x-p)^k.$$

Proof of Theorem 3.1.2. Notice that in fact, the m-th Taylor polynomial of f at a, as well as the remainer associated, depends upon the function f and the point a and $x \in (-r+a, a+r)$. So we shall use the more precise notation

$$P_m(f,a,x) \text{ and } R_m(f,a,x)$$

representing respectively the Taylor polynomial and the remainder (namely $R_m(f,a,x) = f(x) - P_m(f,a,x)$).

Take a real number b satifying $b > a$ and $b - a < r$. Then we shall first prove the following statement:

S$(n+1)$: *For any integer n with $n+1 \geq 0$ and any function $f \in C^{n+1}(a-r, a+r)$, the inequality*

$$\begin{aligned} &\frac{m}{(n+1)!}(x-a)^{n+1} \\ &\qquad \leq R_{n+1}(f,a,x) \\ &\qquad\qquad \leq \frac{M}{(n+1)!}(x-a)^{n+1} \end{aligned} \tag{3.3}$$

holds, whenever $m \leq f^{(n+1)}(x) \leq M$, for any $x \in [a,b]$.

The demonstration is by induction on $n \geq -1$.

If $n = -1$, then (3.3) is the condition $m \leq f^{(0)}(x) = f(x) \leq M$ itself. So it holds trivially.

Assuming that **S**(n) holds, we now show **S**$(n+1)$. Notice that

$$\begin{aligned} \big(P_n(f,a,x)\big)' &= \sum_{k=1}^{n} \frac{f^{(k)}}{(k-1)!}(x-a)^{k-1} \\ &= \sum_{k=0}^{n-1} \frac{f^{(k+1)}}{k!}(x-a)^k \\ &= P_{n-1}(f',a,x) \end{aligned}$$

Therefore, for any $x \in [a, b]$,

$$\begin{aligned} R_n(f', a, x) &= f'(x) - P_n(f', a, x) \\ &= f'(x) - \big(P_{n+1}(f, a, x)\big)' \\ &= \big(R_{n+1}(f, a, x)\big)'. \end{aligned}$$

Now, $\mathbf{S}(n)$ for f' implies that, on $[a, b]$,

$$\frac{m}{n!}(x-a)^n \le R_n(f', a, x) \le \frac{M}{n!}(x-a)^n$$

which is the same as

$$\Big(\frac{m}{(n+1)!}(x-a)^{n+1}\Big)' \le \big(R_{n+1}(f, a, x)\big)' \le \Big(\frac{M}{(n+1)!}(x-a)^{n+1}\Big)'.$$

Since the values of $\frac{m}{(n+1)!}(x-a)^{n+1}$, $R_{n+1}(f, a, x)$ and $\frac{M}{(n+1)!}(x-a)^{n+1}$ coincide at $x = a$, the mean value theorem implies that

$$\frac{m}{(n+1)!}(x-a)^{n+1} \le R_{n+1}(f, a, x) \le \frac{M}{(n+1)!}(x-a)^{n+1}$$

for any $x \in [a, b]$. This completes the induction arguments and hence the proof of $\mathbf{S}(n+1)$ for any integer $n \ge -1$.

If we take $m = \min\{f(x)\colon a \le x \le b\}$ and $m \max\{f(x)\colon a \le x \le b\}$, then the conclusion of $\mathbf{S}(n+1)$ for f implies that

$$m \le R_{n+1}(f, a, x)\frac{(n+1)}{(x-a)^{n+1}} \le M$$

for any fixed value of $x \in (a, b]$. Since $f^{(n+1)}$ is a continuous function on $[a, b]$, the intermediate value theorem[2] implies that there exists some $\xi \in [a, b]$ (depending upon x) such that

$$f^{(n+1)}(\xi) = R_{n+1}(f, a, x)\frac{(n+1)!}{(x-a)^{n+1}}.$$

This proves Theorem 3.1.2 for the case when $x \in [a, a + r)$.

Since the proof of the remaining case when $x \in (a - r, a]$ is similar, we omit it. Therefore, the proof is complete.[3] □

[2]The intermediate theorem: *If a function* $f\colon [a, b] \to \mathbb{R}$ *is continuous and if* $\min_{x\in[a,b]} f(x) \le c \le \max_{x\in[a,b]} f(x)$, *then there exists a point* $\xi \in [a, b]$ *such that* $f(\xi) = c$.

[3]This clever arguments are due to Bary-Soroker and Leher, *On the remainer in the Taylor theorem*, arXiv:0801.1271v2, though the continuity assumption for $f^{(n+1)}$ on $[a, b]$ seem missing in their statement.

3.2 Approximation and estimation of error

The Taylor expansion has many useful features. Let us find out some use by examples.

Can one find the value of $\cos 0.1$? Of course finding the precise value for $\cos\alpha$ for general angle α is known to be an impossible task. Then can one find a close enough approximate value for $\cos 0.1$?

For this purpose, put $f(x) = \cos x$. Then we know $f(0) = \cos 0 = 1$. Also, $f'(0) = -\sin 0 = 0, f''(0) = -1, f'''(0) = 0, f^{(4)}(0) = 1$. Since $f^{(4)}(x) = \cos x = f(x)$, we realize that we know all the Taylor coefficients from $f^{(2k)}(0) = 0$ and $f^{(2k+1)}(0) = (-1)^k$ for every nonnegative integer k. So the $2k$-th order Taylor expansion of $\cos x$ at 0 is obtained as

$$\cos x = 1 + \frac{(-1)}{2!}x^2 + \frac{(-1)^2}{4!}x^4 + \cdots + \frac{(-1)^k}{(2k)!}x^{2k} + R_{2k}$$

where

$$R_{2k} = \frac{(-1)^{k+1}\cos c}{(2k+2)!}x^{2k+2}.$$

(Notice that the remainder may be taken as $\frac{(-1)^{k+1}\sin\tilde{c}}{(2k+1)!}x^{2k+1}$. But since the term after the degree $2k$ term is zero, we may take the remainder after that. And, there is a reason why we prefer this way. Can the reader find it out?)

Now, this consideration implies that with some designated k, we may take the approximate value for $\cos 0.1$ by

$$\cos 0.1 \sim 1 - \frac{1}{2}(0.1)^2 + \cdots + \frac{(-1)^k}{(2k)!}(0.1)^{2k}$$

which is easy to compute. For instance if $k = 2$, then the approximate value is

$$\cos 0.1 \sim 1 - \frac{1}{200} + \frac{1}{240000} = 0.995004166.$$

Well, that is nice, but then, how close is it to the true value of $\cos 0.1$, which is impossible to find?

The Taylor expansion says that the error is

$$R_{2k}(0.1) = \pm\cos c \cdot \frac{1}{6!}(0.1)^6.$$

We do not know the exact value of this error (if we did, we should know the precise value for cos 0.1 also!), but we know that

$$\begin{aligned} |R_{2k}(0.1)| &= |\cos c| \cdot \frac{1}{6!}(0.1)^6 \\ &\leq \frac{1}{720,000,000} < 10^{-8} \end{aligned}$$

which means that the approximate value we obtained from this is accurate up to the 7th decimal below point.

The reader should be able to expand from this example to various cases. This method is also useful with the function $\sqrt[n]{a+x}$; one can see how the table for the (approximate values of) multiple roots were made.

It is also worth noting that one can also resolve the following.

Project: Let f be a real analytic function with $f(a), f'(a), \cdots$ known, so that the radius R of convergence of the Taylor series $T_a f$ is also known. Specify a small positive value r (for the allowed range of error). Let $0 < |s| < R$ be given. Now, can one first know how many terms are necessary in the Taylor series, so that the error for finding the approximate $f(a+s)$ is not more than r? Once the number of terms to be considered is decided, finding the approximate value for $f(a+s)$ is going to be a simple task.

3.3 Taylor series and real-analytic functions

In case a function $f\colon (a-r, a+r) \to \mathbb{R}$ is of class C^∞, i. e., f is of class C^k for every positive integer k, f can have the Taylor expansion of all degrees. In that case the power series

$$[T_a f](x) := \sum_{n=0}^{\infty} \frac{1}{n!} \cdot \left.\frac{d^n f}{dx^n}\right|_a \cdot (x-a)^n$$

arises naturally. This is called the *Taylor series of f at a*.

In many cases, it is convenient if the radius of convergence of the Taylor series $T_a f$ is positive (or, at this occasion, larger than r above). If $[T_a f](x) = f(x)$ for every $x \in (a-r, a+r)$ in addition, then calculus on f can enjoy many convenient features. Of course, not all functions are so privileged.

A typical well-known example is as follows: Let

$$f(x) = \begin{cases} e^{-\frac{1}{x}} & \text{if } x > 0 \\ 0 & \text{if } x \le 0. \end{cases}$$

Then f is of class C^∞ at every real number value of x. Moreover, $f^{(k)}(0) = 0$ for any positive integer k. Therefore its Taylor series $T_0 f = 0$ identically. But $f(x) > 0$ for any $x > 0$. So this function f cannot coincide with its Taylor series $T_0 f$.

Definition 3.3.1 A real-valued function f is called *real-analytic* at a if f is defined on an open interval say $(a-r, a+r)$ for some $r > 0$ and satisfies the following properties:

(1) f is of class C^∞.

(2) The Taylor series $T_a f$ has positive radius of convergence, say $\delta > 0$.

(3) $f(x) = [T_a f](x)$ for any x with $|x-a| < \delta$.

In general, a function is said to be *real analytic* if it is real analytic at every point of its domain of definition.

The example above is not real analytic at 0. However, it is real analytic if the domain of definition is taken as $D = \mathbb{R} \setminus \{0\}$. We choose not to discuss, here, any further on the real analyticity.

Example 3.3.1 Many functions familiar to us are real analytic.[4] All polynomial functions, and all rational functions except at the points where the function is undefined are real analytic. $e^x, \ln x$ are real analytic. $\sin x, \cos x$ are real analytic. The sum, product and quotients of real analytic functions are real analytic. Composition of real analytic functions are real analytic. The notation for the class of real analytic functions is C^ω, usually.

Example 3.3.2 Find the Taylor expansion of the following functions at the appropriate center point (say at 0):

1. $y = e^x$.
2. $y = 9^x$.
3. $y = \sin x$.
4. $y = \cos x$.
5. $y = \sqrt[3]{1+x}$.
6. $y = \tan x$.
7. $y = \ln(1+x)$.

All these functions are real analytic near the origin. Can you verify it?

[4]The terminology *real analytic* may not appeal to the reader, I suppose. Traditionally, the term *analytic function* was overwhelmingly used for the complex analytic functions. During the period when the study of complex analytic functions swept through the mathematical world, even the terminology "Function theory" just meant the theory of complex analytic functions and applications! That "madness" has calmed down now, but still the traces remain in many places.

Chapter 4

Some differential equations

Perhaps the most famous case of application of Calculus may be by way of differential equations. This particular study is so useful in many sciences and engineering that university mathematics curricula offer several courses on various levels. Here, however, we shall learn only a few basic examples, theories and techniques.

4.1 Separable equations

4.1.1 Growth of bacteria

At a given moment, say $t = t_0$, let us say that the number of bacteria (in a given region) was $P(t_0)$. Then one of the interesting problems was to know what $P(t)$, the number/population of bacteria, would be at time t.

Why would it be of an interest? It is well known that many epidemics are caused and governed by bacteria and their growth in population. This could be an explanation enough.

Then, how would one find the function $P(t)$?

The nature of bacteria has been studied by much effort in the laboratory. Their study has revealed that the rate of growth of the population of bacteria in a short term is proportional to the population of initial type. And that proportion (namely, the ratio) is almost constant, for some time. In a mathematical formulation, it becomes

$$\frac{P(t+h)-P(t)}{h} \propto P(t), \quad \forall h \text{ with } |h| \text{ small.} \tag{4.1}$$

So for the time being, the equation to consider becomes

$$P'(t) = kP(t), \tag{4.2}$$

where k is some constant, which depends upon the nature of the bacteria under consideration.

Please keep in mind that what we would like to understand is the function $P(t)$, which represents the population of bacteria at time t. [In application, more interest is given to $P(t)$ at the future time points t.] If $P(t)$ becomes considerably small, for instance, then the epidemic would calm down!

Then what are the questions to which one would like the answers? Some of them are as follows:

- When will the value of $P(t)$ be bounded by a certain (safe) number?
- When will the value of $P(t)$ reach at the critical dangerous level?
- What should be done to avoid the impending crisis?

All these are related to (1) the understanding of the function $P(t)$ satisfying the given "governing equation" for $P(t)$, and (2) can one modify the equation (so that the new factors be considered to change the equation) to reach at more precise a function $P(t)$, for instance. In any situation, the key is in understanding the function $P(t)$.

4.1.2 Differential equations

Notice that the problem therefore is the following:

Step 1: From the problem(s) of nature, derive the equation(s) that the critical factor function and their derivatives satisfy. Such equations are called the differential equation.

Step 2: Solve for the function in the given equation.

Then in an application, one has to analyze the solution function for their purposes. But in mathematics, it is reasonable to concentrate upon the first two steps. If we rewrite the discussion of preceding section, the differential equation to consider is

$$y' = ky \tag{4.3}$$

and the aim should be **finding the function** $y = f(t)$ **of time variable** t.

4.1.3 Solving for y in a separable equation

Let us start with $y' = ky$. Except when $y = f(t)$ takes the value 0, the equation can be converted to

$$\frac{1}{y} \cdot y' = k.$$

This implies that

$$\int_{t_0}^{t} \frac{1}{f(t)} \cdot f'(t)dt = \int_{t_0}^{t} kdt.$$

Then

$$\ln |f(t)| - \ln |f(t_0)| = k(t - t_0)$$

which implies

$$f(t) = Ce^{k(t-t_0)},$$

for some constant C. This could be just one function, or two functions. But at least we have obtained some significant result. Now the following question arises: how can one deal with C and k?

If we knew $f(t_0) = P(t_0)$, then we see that $C = P(t_0)$. If we knew one more information such as $f(t_1)$ with $t_1 \neq t_0$ we should be able to compute the value of k. Through this process, we can find $f(t)$, the solution function, and achieve the goal.

Reviewing this process, the following symbolic computation scheme can be formulated. Now, we shall use x instead of t, and put $t_0 = 0$.

$$\begin{aligned}
y' &= ky \\
\frac{dy}{dx} &= ky \\
\frac{1}{y}dy &= k\ dx \\
\int \frac{1}{y}dy &= \int kdx \\
\ln|y| &= kx + C.
\end{aligned}$$

This last implies $y = \pm e^C e^{kx}$, which we rewrite as $y = Ce^{kx}$.

In case the differential equation can be written symbolically as

$$P(y)dy = Q(x)dx$$

then the above procedure shows that one may obtain

$$\int P(y)dy = \int Q(x)dx$$

which does not contain derivatives any more. This at least gives the relation between y and x, thus the solution y is "given" implicitly as an implicit function of variable $x(=t)$. Such differential equations are called *separable equations*.

Example 4.1.1 The population model $y' = ky$ with $k > 0$ is too simple. In fact, if such growth continues, soon there will be too many bacteria and they will compete with others. In fact when they are pressured by the others, they tend to produce some material for others to experience difficulty in surviving (such as antibiotics). With such, the equation changes to

$$y' = ky - ry^2$$

with $k > 0$ and $r > 0$.

Solution. Then symbolic computation gives:

$$\int \frac{1}{ky - ry^2}dy = \int dx$$

This is another separable equation. In order to resolve the left hand side, one needs partial fraction technique (Section 1.3.2). The remaining detail is left for the reader. ♣

4.1.4 General solution, Special solution, Initial conditions

The differential equation of explicit/implicit type

$$y' = F(x, y) \text{ or } G(x, y, y') = 0$$

usually produces a formula for $y = f(x)$ with a constant undecided, as the key step is the indefinite integral which almost always brings in the undetermined constant. In order to fix this problem, the above problem comes with additional information such as

$$y(0) = y_0,$$

where y_0 is an explicit value. Thus the problem can be written in a concise form

$$\begin{cases} y' = F(x, y) \text{ or } G(x, y, y') = 0 \\ \quad y(0) = y_0. \end{cases}$$

This is called the *initial value problem*. For example,

$$\begin{cases} y' = xy \\ \quad y(0) = 2. \end{cases}$$

Notice that this separable equation gives a unique solution function.

The solution with the undecided constant is usually called the *general solution* as it contains (almost) all possible solutions of the differential equation. With constant decided, the solution is usually called a *particular* solution. The solution to an initial value problem is therefore a particular solution.

4.2 First order linear equations

The next type of differential equation is

$$y' + p(x)y = q(x), \tag{4.4}$$

where $p(x), q(x)$ are functions of x. A good example is

$$y' - 2y = x. \tag{4.5}$$

Notice that this is not a separable equation. Since the equation contains only the first order terms of y, y', such equation is called a *first order linear equation*. Then how can one solve this equation for y?

Solution method: If $(uy)' = u(x)(y'+p(x)y)$, for some smooth function $u(x)$, then the equation (4.4) becomes

$$(uy)' = u(x)q(x).$$

Therefore, we get

$$\begin{aligned} u(x)y &= \int u(x)q(x)dx \\ y &= \frac{1}{u(x)}\int u(x)q(x)dx \end{aligned}$$

which leads us to the general solution.

Thus, we need to obtain $u(x)$. By a direct computation it suffices that $u(x)$ is a function satisfying the equation $u' = p(x)u$. This is a sparable equation and a solution is

$$u(x) = e^{\int p(x)dx}.$$

(Note that one particular solution suffices.) Altogether, one gets a general solution. (Then initial value problem can be resoloved accordingly.) Such u, which makes the machinery work, is called the *integrating factor.*

Let us do the explicit initial value problem $y' - 2y = x, y(0) = 1$. The integrating factor is

$$u(x) = e^{\int(-2)dx} = e^{-2x}.$$

(We just need some u, not all possible u.) Thus the equation becomes

$$(e^{-2x}y)' = xe^{-2x}.$$

This can be solved as follows:

$$e^{-2x}y = \int xe^{-2x}dx$$
$$y = -\frac{x}{2} - \frac{1}{4} + Ce^{2x}$$

which gives the general solution. With initial value $y(0) = 1$, one gets

$$1 = -\frac{1}{4} + C$$

and consequently $C = \frac{5}{4}$. So the solution function for the initial value problem is

$$y = -\frac{x}{2} - \frac{1}{4} + \frac{5}{4}e^{2x}.$$

A primary example of first order linear equation is so called "mixing problem" as in the following

Example 4.2.1 A pond contains 20 million liters of water with a certain pollution material of 20 tons. 1 million liters of water flows in every time t (measured in years) with $2\sin t$ tons of the same pollution material, gets mixed in instantly, and then flows out at the same rate (i. e., 1 million liters per year). Find the formula for the amount of pollution material at time t (years). Can one estimate the time required for the amount (in tons) of the pollution content in the pond to be lowered to an amount less than 10 tons?

Solution. Let $q(t)$ be the amount of pollution material in the pond at time t. Then

$$\begin{aligned} q'(t) &= \text{rate of input (ton/year)} - \text{rate of output (ton/year)} \\ &= 2\sin t - \frac{q(t)}{20 \times 10^6} \cdot 10^6, \end{aligned}$$

which is

$$q'(t) + \frac{1}{20}q(t) = 2\sin t.$$

The solution method introduced yields

$$\begin{aligned}(e^{\frac{t}{20}}q(t))' &= e^{\frac{t}{20}}\sin t \\ q(t) &= e^{-\frac{t}{20}}\int e^{\frac{t}{20}}\sin t\ dt\end{aligned}$$

With $q(0) = 20$ (tons), one can compute an explicit formula for $q(t)$. To answer the time required for the amount (in tons) of the pollution content in the pond to be lowered to the level under 10 tons, one has to solve for the value of t in

$$q(t) < 10.$$

We leave checking the detail to the reader. ♣

4.3 Solution by series

In the case that the differential equation with initial condition is given by

$$y' = p(x)y^2 + q(x), \; y(0) = 1,$$

the preceding techniques cannot help us much in finding the solution y. In general, the differential equations which can be explicitly solvable are not many!

But if $p(x)$ and $q(x)$ are real analytic in addition, then the celebrated Cauchy-Kowalewsky theorem[1] says that the solution exists and is also real analytic.

Example 4.3.1 Solve for y in the initial value problem $y' - x^2 y = e^x$, $y(0) = 0$, using series.

Solution. Let $y = \sum_{n=0}^{\infty} a_n x^n$, then

$$y' = \sum_{n=1}^{\infty} n a_n x^{n-1} = \sum_{n=0}^{\infty} (n+1) a_{n+1} x^n.$$

Then the initial condition $y(0) = 0$ means $a_0 = y(0) = 0$. The differential equation becomes:

$$\sum_{n=0}^{\infty} (n+1) a_{n+1} x^n - \sum_{n=0}^{\infty} a_n x^{n+2} = \sum_{n=0}^{\infty} \frac{1}{n!} x^n,$$

which is equivalent to

$$a_1 + 2a_2 x \sum_{n=0}^{\infty} [(n+3)a_{n+3} - a_n] x^{n+2} = 1 + x + \sum_{n=0}^{\infty} \frac{1}{(n+2)!} x^{n+2}.$$

Therefore,

$$a_0 = 0, a_1 = 1, a_2 = \frac{1}{2},$$

and

$$(n+3)a_{n+3} - a_n = \frac{1}{(n+2)!}$$

for every $n = 0, 1, 2, \ldots$. This will produce (in theory) all values for a_n, for $n = 0, 1, 2, \ldots$. Therefore (in theory) we find the solution $y = \sum a_n x^n$ represented by a power series. □

[1] It was one of the great achievements by Sof'ya Kowalewskaya (1850-1891).

Remark 4.3.1 For the nonlinear equation $y' = p(x)y^2 + q(x)$ with initial condition $y(0) = 1$, finding the series solution is difficult, as one experience difficulty in expressing $\left(\sum a_n x^n\right)^2$ by a power series, as the coefficients are complicated. But this gives solution to some order, which already carries some meaning in application.

4.4 Difference equations and iteration

The series solution method introduced in the preceding section is impressive, but it is not impressive enough when it was applied to the first order linear equation. (Later, when the student may encounter the second order linear equation with real analytic function coefficients, there the series method will show its real power.) And yet, there is another approximation method that is widely used to obtain approximate solutions to a differential equation.

Take the example of

$$y' = xy^2 + e^x, \quad y(0) = 1.$$

for instance. This equation is not even linear, and not separable. In fact this is not easy to find the exact solution by any known means. Then use

$$y' = f'(t) = \lim_{h\to 0} \frac{f(t+h) - f(t)}{h} \sim \frac{f(t+h) - f(t)}{h}$$

if $|h|$ is sufficiently small. The information we start with are therefore

$$\begin{aligned} f(0) &= 1 \\ f(t+h) &\sim f(t) + ht(f(t))^2 + he^t. \end{aligned}$$

From these, we immediately obtain

$$\begin{aligned} f(h) &= f(0+h) \sim f(0) + 0 + he^0 = 1 + h. \\ f(2h) &= f(h+h) \\ &\sim f(h) + h^2 f(h)^2 + he^h \\ &= 1 + h + h^2(1+h)^2 + he^h, \end{aligned}$$

and so forth.

Inductively, one may obtain all values of $f(mh)$ for every integer m for any small h, approximately. Of course this method does not give any particular value of $f(x)$ and worse, the values $f(mh)$ are obtained only approximately. Finding reasonable estimate for the error is also in order. But this at least shows a possibility while the precise solution to the differential equation is not found.

Chapter 5

Calculus proper

So far, we have been concentrating upon the technical side of theory of functions by way of Calculus, differentiation and integration. While doing those, it was obvious that the necessity of making the foundation of knowledge as solid as possible emerged as more and more important an issue. I understand that the reader is most likely not going to pursue a career in pure mathematics. But still laying a solid foundation of Calculus have proved to be very much fruitful in doing whichever science or engineering.

5.1 Functions

We shall use the expression $f(x)$ representing the operation which assigns a unique real number value $f(x)$ to each real value x belonging to a specific set, say A, of real numbers. Such $f(x)$ is typically given by certain formula such as $f(x) = x^2 + 1$ for instance. In such a case $f(3) = 3^2 + 1 = 10$, $f(-10) = (-10)^2 + 1 = 101$ etc. For this formula, the set A may be the set of all real numbers, and one says in mathematics that such $f(x)$ defines a *function*. The concept is defined as follows:

Definition 5.1.1 Let A, B be sets and let f be a subset of the set of ordered pairs $A \times B := \{(a, b) \colon a \in A,\ b \in B\}$. Then, f is said to be a *function* if it satisfies the following two conditions:

(1) For any $a \in A$, there exists $b \in B$ such that $(a, b) \in f$.

(2) If $(a, b_1) \in f$ and $(a, b_2) \in f$, then $b_1 = b_2$.

In such a case, emphasizing the fact that the function is a special correspondence, we denote the function by the notation $f\colon A \to B$.

This definition may be somewhat abstract, but it is actually the object that we are familiar with. Let us use the notation

$$b = f(a)$$

meaning $(a, b) \in f$. Going back to the explanation of the beginning of this section, let us consider the (special) case when A and B are subsets of the set of real numbers (or, the real line). There f was a formula $f(x)$ and when x is replace by a number a, the $b = f(a)$ means, in our "familiar terms", the output obtained by calculating the value of the formula is b. By the abstract definition above, it means $(a, b) \in f$, of course.

In this regard, two conditions in Definition 5.1.1 simply mean two features: First, for any *input* (for instance $a \in A$ there must exists an element $b \in B$, the *output*. The second feature is that the result of the calculation by the formula, i. e., the output b has to be unique. If the output displays more than two distinct values or, if there is no output at all, then f is not a function.

The reader should be familiar with $f(x) = x^2$. This is of course a function, very typical. In the formal definition (Definition 5.1.1), if we denote by $\mathbb{R}$ the set of real numbers, then

$$f = \{(x, y) \in \mathbb{R} \times \mathbb{R}\colon y = x^2\}.$$

Thus the concepts of the function and its graph, discussed conventionally, are the same, although the former is understood as conceptual, and the latter is regarded as geometric.

5.2 Limits

For many purposes, the concept of "x approaching a" in $\mathbb{R}$ has been discussed. Of course, this should mean intuitively that the value of x will move closer and closer to the fixed real number a. This can be explained by the "logical" phrase

"for any x with $|x - a| < \delta$ (at each time when a $\delta > 0$ is given)"

In order to develop some feelings towards this logical statement which includes the quantifier terminology such as *for any*, let us look at the following:

Lemma 5.2.1 *If a is a real number, and if $|a| < \delta$ for any $\delta > 0$, then $a = 0$.*

Proof. Suppose that $a \neq 0$. Then $|a| > 0$. Of course

$$|a| - \frac{|a|}{2} = \frac{|a|}{2} > 0.$$

Take $\delta = \frac{|a|}{2}$ then the assumption that $|a| < \delta$ for any $\delta > 0$ does not hold, which is a contradiction. This completes the proof. □

Then we shall give the definition of the sentence (i. e., an expression)

$$\lim_{x \to a} f(x) = L, \tag{5.1}$$

where a and L are real numbers and where f is a function defined on an interval containing a. This expression is understood intuitively as follows:

The value of $f(x)$ converges to L, as x approaches a.

But this sentence is not exactly precise, nor logical. Also mathematicians would like to exclude the case $x = a$, because that is not only away from the principal interest of discussion on this "limit" concept but also misleading in many cases. Keeping the preceding arguments in mind let us learn the precise definition of (5.1).

Definition 5.2.1 Let a and L be real numbers and let f a function defined on an interval I containing a. By

$$\lim_{x \to a} f(x) = L, \tag{5.2}$$

we mean the following: *For any $\epsilon > 0$, there exists $\delta > 0$ satisfying that*

$$|f(x) - L| < \epsilon$$

for every $x \in I$ with $0 < |x - a| < \delta$.

Over time, I learned from my own experiences that students did not find this concept so easy to understand. But the definition actually makes sense, even intuitionally. Let us give a try.

First take a positive number, say 100, and let ϵ be that number. Then the condition $|f(x) - L| < \epsilon$ says that there is a number, say 10 (which is δ), such that, whenever $|x - a|$ is smaller than 10, the value $f(x)$ is between $L - 100$ and $L + 100$. This is fine, but that does not seem to do much towards the convergence.

Then take a different number, say 0.1, which is the value of ϵ this time. The definition of limit says that, even so, one can still find a value for δ, say 0.01, such that, whenever x is between $a - 0.01$ and $a + 0.01$, the value of $f(x)$ satisfies $|f(x) - L| < 0.1$. If one accepts that the second case is a better approximation of their intuitive understanding of the limit, then one may repeat the similar arguments varying the values of $\epsilon > 0$, and the values of $\delta > 0$ subordinate to the values of ϵ.

Does this mean something to the reader? I hope so; the keys are hidden in the quantifiers such as "for any" and "for some".

In passing, I would like to point out that the condition $0 < |x - a|$ is the same as $x \neq a$. Namely, in the discussion of the limit of $f(x)$ as x approaching a, the value $f(a)$ is not important, and hence is natural to be precluded.

Let us now look at a few examples and learn how to use Definition 5.2.1.

Example 5.2.1 Show that

$$\lim_{z \to 1} \frac{1 - x^2}{1 - x} = 2.$$

Proof (Conventional). Let $\epsilon > 0$ be given arbitrarily. Then we will have to find $\delta > 0$ that verifies the condition that defines the limit definition. So start with

$$\left|\frac{1-x^2}{1-x} - 2\right| < \epsilon.$$

Notice that the definition comes with the condition that $x \neq 1$. So the inequality above is the same as

$$|1 + x - 2| < \epsilon$$

which is equivalent to $|x - 1| < \epsilon$. Therefore we let $\delta = \epsilon$. Then the proof is complete. □

This is accepted as the proof, by most mathematicians. But logically speaking, the formal proof should be organized differently.

Proof (Logical). Let $\epsilon > 0$ be given (arbitrarily). Then choose δ by $\delta = \epsilon$. Then for any x with $0 < |x - 1| < \delta$, it follows that

$$\left|\frac{1-x^2}{1-x} - 2\right| = |1 + x - 2| = |x - 1| < \delta = \epsilon.$$

Therefore, the proof is complete. □

The logical proof did the following: With $\epsilon > 0$ arbitrarily chosen, the positive constant δ subject to ϵ was then selected (to be the same as ϵ in this case). Then the rest of argument shows that such a choice of δ verifies the inequality

$$\left|\frac{1-x^2}{1-x} - 2\right| < \epsilon$$

for any x satisfying $0 < |x - 1| < \delta$, as required for the proof of

$$\lim_{x\to 1} \frac{1-x^2}{1-x} = 2.$$

This proof is logically precise; there is no question about that. On the other hand, how did the writer of this proof know to take δ as such?

Yes, you knew the answer. It was possible, by the arguments of the first proof (conventional). Also it is almost (if not already fully) obvious that the logically correct proof will follow as soon as the conventional proof is constructed. This is the reason why many mathematicians simply write the conventional proof and omit the logical proof.

Theorem 5.2.2 *Let f and g be real valued functions defined on a subset A of $\mathbb{R}$ with $a \in A$, and let C be a real number. Then the following identities hold:*

(1) $\lim_{x \to a} Cf(x) = C \lim_{x \to a} f(x)$,

(2) $\lim_{x \to a} \big(f(x) + g(x)\big) = \lim_{x \to a} f(x) + \lim_{x \to a} g(x)$,

(3) $\lim_{x \to a} f(x)g(x) = \lim_{x \to a} f(x) \lim_{x \to a} g(x)$,

(4) $\lim_{x \to a} \dfrac{f(x)}{g(x)} = \dfrac{\lim_{x \to a} f(x)}{\lim_{x \to a} g(x)}$,

provided that the limits constituting the right hand sides of the identities exist (as finite values) and, additionally for (4), $\lim_{x \to a} g(x)$ *is not zero.*

The deal is that proving this theorem requires the definition of the limit as illustrated in the examples above, but once Theorem 5.2.2 is established, one need not always go back to the rigorous arguments using the ϵ-δ definition (Definition 5.2.1).

5.3 Continuity

Let us set some notation. Let a, b be real numbers with $a \leq b$.

$$\begin{aligned}(a,b) &:= \{x \in \mathbb{R}\colon a < x < b\}\\(a,b] &:= \{x \in \mathbb{R}\colon a < x \leq b\}\\[a,b) &:= \{x \in \mathbb{R}\colon a \leq x < b\}\\[a,b] &:= \{x \in \mathbb{R}\colon a \leq x \leq b\}.\end{aligned}$$

Of course, if $a = b$, then the first three sets are empty. The first one is called an *open interval*, and the last one a *closed interval.* The second and third are usually called the *half-open* intervals. Any one of these are just called intervals.

Let A be a subset of the real line $\mathbb{R}$. Emphasizing that a function $f \subset A \times \mathbb{R}$ assigns the real value $f(a) \in \mathbb{R}$ to the element $a \in A$, mathematicians sometimes denote the function f by $f\colon A \to \mathbb{R}$. But this is really the same as saying that the function $y = f(x)$ with its domain of definition A. We shall use these (mutually equivalent) notation whichever is more appropriate or convenient.

Definition 5.3.1 Let $f\colon A \to \mathbb{R}$ be a function. Let $a \in A$. Then f is said to be *continuous at a* if

$$\lim_{x \to a} f(x) = f(a),$$

which means that, for any $\epsilon > 0$, there exists a positive number δ such that $|f(x) - f(a)| < \epsilon$ for any $x \in A$ satisfying $|x - a| < \delta$.

Theorem 5.3.1 *If the functions $f, g\colon A \to \mathbb{R}$ are continuous at $a \in A$, then so are $f \pm g$, cf, fg and f/g, provided that, especially for f/g, $g(a) \neq 0$.*

5.4 Differentiation

Start with the following working definition.

Definition 5.4.1 A subset A of the real line $\mathbb{R}$ is said to be *open* if A is a union of open intervals.

Notice that if A is open, then for any $p \in A$, there is an open interval $(a, b) \subset A$ with $p \in (a, b)$, and hence $a < p < b$. Therefore, there is $r > 0$ (depending upon p and A) such that $(-r + p, p + r) \subset A$.

Definition 5.4.2 For a function $f\colon A \to \mathbb{R}$ and a point $p \in A$, we define the *Newton quotient* of f at p is defined by the expression

$$\lim_{x \to p} \frac{f(x) - f(p)}{x - p}.$$

If this limit exists, then it is called the *derivative of f at p* and is denoted by $f'(p)$. Furthermore, in such a case, the function f is called *differentiable* at p.

In case the function is differentiable at every point of A, then f is called *differentiable on A*. There are many differentiable functions, even on the whole real line. All polynomial functions, the sine and cosine functions as well as the exponential function are such functions. The rational functions are also differentiable at every point where the denominator is nonzero.

Remark 5.4.1 If we put $x = a + h$ in the formula for the Newton quotient above, we see immediately that

$$\lim_{x \to a} \frac{f(x) - f(a)}{x - a} = \lim_{h \to 0} \frac{f(a + h) - f(a)}{h}.$$

Of course one has to establish (i. e., prove) that

$$\lim_{x \to a} h = 0 \Leftrightarrow \lim_{h \to 0}(x - a) = 0,$$

but, it is obvious. (If one insists that a formal proof is necessary, we leave it to the reader.)

5.4.1 Linearization

Let $y = f(x)$ be a real-valued function defined on an open subset A of the real line $\mathbb{R}$. If f is differentiable at every point of the interval $(-r+p, r+p)$ for some constant $r > 0$ with $I := (-r+p, r+p) \subset A$ so that the derivative $y = f'(x)$ is a well-defined continuous function on I. Then for any $u \in I$ we observe that

$$\begin{aligned} 0 &= \lim_{x \to u} \frac{f(x) - f(u)}{x - u} - f'(u) \\ &= \lim_{x \to u} \frac{f(x) - \big(f(u) + f'(u)(x-u)\big)}{x - u}. \end{aligned}$$

If we denote by

$$y = L_u(x) := f(u) + f'(u)(x - u)$$

then the function $y = L_u(x)$ represents the straight line in the xy-plane with slope $f'(u)$ passing through the point $(u, f(u))$.

Moreover, if we put

$$E_u(x) = |f(x) - L_u(x)|,$$

then

$$\lim_{x \to u} \frac{E_u(x)}{|x - a|} = 0.$$

In function theory the following definition is popular.

Definition 5.4.3 (The classes $o_a(f), O_a(f)$) If f and g are functions defined on an open interval containing $a \in \mathbb{R}$, then f is said to *belong to the class* $o_a(g)$, i. e., in notation $f \in o_a(g(x))$, if

$$\lim_{x \to a} \frac{f(x)}{g(x)} = 0.$$

Moreover, if there is a constant $C > 0$ such that there exists $r > 0$ such that

$$|f(x)| \leq C|g(x)|, \quad \forall x \text{ with } |x - a| < r$$

then we denote by

$$f \in O_a(g).$$

This defines the class $O_a(g)$.

Traditionally, mathematicians wrote $f(x) = O_a(g(x))$, or even $f(x) = O(g(x))$ as $x \to a$. But since the right-hand sides represent a class of functions, they must have thought that the equality notation should not confuse the reader.[1]

This definition of classes represent the order of "growth" of the function $f(x)$ near the point a in comparison with the growth of $g(x)$ near the point a. In word, $f \in o_a(g)$ (or, $f = o_a(g)$) means that f vanishes to 0 *faster* than the function g does, as x approaches a. [Likewise, $f \in O_a(g)$ means that the growth of f is not any more than that of g as x approaches a. In this viewpoint, we were saying that the difference $E_u(x)$ between the values of the function $f(x)$ and its *linearization* $L_u(x)$ belongs to the class $o_a(|x-a|)$.

Proposition 5.4.1 *Let $m_1, m_2, a, b \in \mathbb{R}$ and $\ell_1(x) = m_1(x-a)+b$ and $\ell_2(x) = m_2(x-a)+b$. If $\ell_1 - \ell_2 \in o_a(x-a)$, then $\ell_1 = \ell_2$ as functions.*

Proof. It suffices to show that $m_1 = m_2$. Since $\ell_1(x) - \ell_2(x) = (m_1 - m_2)(x - a)$, we see that

$$0 = \lim_{x \to a} \frac{\ell_1(x) - \ell_2(x)}{|x-a|} = m_1 - m_2$$

by the assumption. This completes the proof. □

Now, the meaning of the linearization L_u can be discussed geometrically. First, $y = L_u(x)$ represents the straight line passing through the point $(u, f(u))$ with slope $f'(u)$. The property that $L_u \in o_u(|x-u|)$, in the light of Proposition 5.4.1, approximates f near $x = u$ the best, among straight lines. This equation $y = L_u(x) = f'(u)(x-a) + f(u)$ therefore is rightly called the *tangent line* to $y = f(x)$ at $x = u$. Needless to say, the differentiability of f at u implies the existence of the tangent line L_u, which gives the best approximation to $f(x)$ among straight lines.

5.4.2 Tangent line

We choose to post the last part of discussion from the preceding section again.

Definition 5.4.4 For a function f defined on an open interval containing p, a line $\ell(x) = m(x-p) + f(p)$ is called a *tangent line to $y = f(x)$ at p*, if $f(x) - \ell(x) = o_p(|x|)$.

[1]Tradition, Ah! To follow or not to—that is the question.

Proposition 5.4.2 *A function f defined on an open interval containing p admits a tangent line if, and only if, f is differentiable at p. Moreover, in such a case, the tangent line is unique and is represented by the linearization $L_p(x) = f'(p)(x - p) + f(p)$.*

We leave the proof to the student.

5.5 Smooth functions

5.5.1 Classes $\mathcal{C}^k, \mathcal{C}^\infty$

For the present and future purposes, we discuss the class of functions. A function $y = f(x)$ defined on an open interval A is differentiable at every point, then this function is said to belong to the class $\mathcal{C}'(J)$, or simply $\mathcal{C}'$.

Notice that every differentiable function is continuous. (Exercise: Prove it!) But the derivative may not in general be continuous. Can one find an example? One well-known example is as follows:

Example 5.5.1 Let

$$f(x) = \begin{cases} x^2 \sin \frac{1}{x}, & x > 0 \\ 0, & x \leq 0. \end{cases}$$

This defines a function that is differentiable at every point of the real line $\mathbb{R}$. But its derivative is not continuous at 0. Proving the necessary claims is left to the reader as exercises.

This calls for the concept of another class, slightly more restricted:

$$\mathcal{C}^1(J) := \{f \in \mathcal{C}'(J) \colon f' \text{ is continuous}\}.$$

A function $f \in \mathcal{C}^1(J)$ is usually said to be *continuously differentiable.*

Extension of such consideration is also known and in use. For any positive integer k, f is said to *be of the class* $\mathcal{C}^k$ if f is differentiable at every point k-times, and the k-th derivative $f^{(k)}$ is continuous. If a function f is of class $\mathcal{C}^k$ for every positive integer k, then we say that f is *differentiable indefinitely many times*, and denote by $f \in \mathcal{C}^\infty$.[2]

Of course, the first derivative is written casually by f', and then the second derivative by f'', and f''', but one can see why $f^{(k)}$ emerged as the notation.

Differentiable functions are casually called, in mathematics, *smooth functions.* Moreover, the notation $\mathcal{C}^0$ is also used, for the class of continuous functions.

[2] $f \in \mathcal{C}^\infty$ is casually read "differentiable infinitely many times," also.

5.5.2 Typical smooth functions

Leibniz, one of the first founders of Calculus, used the notation

$$\frac{df}{dx}, \frac{d}{dx}f$$

for the (first) derivative f'. Then further notation such as

$$f^{(n)} = \left(\frac{d}{dx}\right)^n f = \frac{d^n f}{dx^n}$$

are also in use. For the open intervals I, J, consider the functions $f\colon I \to J \subset \mathbb{R}$ and $g\colon J \to \mathbb{R}$, the *composite* function $g \circ f\colon I \to \mathbb{R}$ is defined by

$$[g \circ f](x) = g(f(x)), \quad \forall x \in I.$$

Now from the definition and the properties of limit, one obtains:

Theorem 5.5.1 *For a constant $c \in \mathbb{R}$ and smooth functions f and g defined on an open interval J, the following hold for every $x \in J$.*

(1) $(cf(x))' = cf'(x)$.

(2) $(f(x) + g(x))' = f'(x) + g'(x)$.

(3) $(f(x)g(x))' = f'(x)g(x) + f(x)g(x)$.

(4) $(f(x)/g(x))' = (f'(x)g(x) - f(x)g'(x))/(g(x)^2)$, *whenever* $g(x) \neq 0$.

The proofs are straightforward from Theorems 5.2.2 and 5.3.1. Furthermore, for the composite functions, we have the following.

Theorem 5.5.2 (The Chain Rule) *If I, J are two open intervals, and if $f\colon I \to J \subset \mathbb{R}$ and $g\colon J \to \mathbb{R}$ are smooth functions, then*

$$(f \circ g)'(x) = f'(g(x))g'(x), \quad \forall x \in I.$$

The influence of these two theorems propagates far and wide: first, they imply that all polynomials such as $a_0 + a_1x + \cdots + a_nx^n$ is smooth as soon as one proves that the functions $f(x) = x$ is differentiable, with $f'(x) = 1$. Second, (3) of Theorem 5.5.1 implies that $(x^m)' = mx^{m-1}$ via induction. (1)-(3) of Theorem 5.5.1 yields the differentiability of the polynomials and the formulae for their derivatives. Third, it explains how to differentiate rational functions, the fraction function both the numerator and the denominator of which are polynomials. We leave the verifications and computations to the reader as exercises.

It also tells us how to differentiate the multiple root functions (when defined). Let us take $h(x) = \sqrt{x}$ defined for $x > 0$. From $h(x) \cdot h(x) = x$, Formula (3) of Theorem 5.5.1 yields that

$$1 = h'(x)h(x) + h(x)h'(x) = 2h(x)h'(x)$$

which implies

$$(\sqrt{x})' = \frac{1}{2\sqrt{x}}.$$

Of course, this computation is subject to the differentiability of the function $h(x) := \sqrt{x}$. This is put into the following exercise.

Example 5.5.2 Show, for every $a > 0$, that

$$\lim_{x \to a} \frac{\sqrt{x} - \sqrt{a}}{x - a} = \frac{1}{2\sqrt{a}}.$$

The discussions up to this point tell us that all polynomial functions, rational functions, and the radicals (i. e., the multiple root functions) are smooth functions on the intervals on which they are defined.[3] There are of course many more smooth functions such as trigonometric functions, hyperbolic functions and others (called the transcendental functions). We shall learn them as the exposition progresses.

[3]These functions are traditionally called the *algebraic functions*.

5.6 The Mean-Value Theorem

5.6.1 Mean-Value theorem

Intuitively, for a differentiable function (i. e., a function differentiable at every point of the domain A of definition, which is also an open subset of $\mathbb{R}$) f, the derivative and the tangent lines are good for discussing local maxima and minima.[4] The first result one may think of in this regard is

Proposition 5.6.1 *Let f be differentiable at every point of an open set A in $\mathbb{R}$. If $p \in A$ admits $r > 0$ such that $(-r+p, p+r) \subset A$ and $f(x) \leq f(p)$ for all $x \in (-r+p, r+p)$, or if $f(x) \geq f(p)$ for $x \in (-r+p, r+p)$, then $f'(p) = 0$.*

proof. Let p be such that

$$f(x) \leq f(p), \quad \forall x \in (-r+p, p+r).$$

Assume the contrary that $f'(p) \neq 0$. If $f'(p) = -c < 0$ then, for $\epsilon = \frac{c}{2}$, there exists $\delta > 0$ such that

$$\left|\frac{f(x)-f(p)}{x-p} - f'(p)\right| < \epsilon$$

whenever $|x-p| < \delta$. If $\delta > r$, then this statement remains true for all x with $|x-p| < r$. So we may replace δ by the smaller of δ and r.

Now the inequality above implies that, for any x with $-r < x-p < 0$,

$$f(x) - f(p) < (-\epsilon + f'(p))(x-p) = -\frac{c}{2}(x-p) < 0.$$

In particular, if $x_0 = p - \frac{\delta}{2}$, then $x_0 \in (-r+p, p+r)$ and

$$f(x_0) > f(p),$$

which is a contradiction to the assumption.

Since the proof of the other cases follow by similar arguments, the proof is complete. □

Then we introduce

[4]The word *maximum* (respectively, *minimum*) is of Latin origin, which means the largest (respectively, the smallest) value of the function. $f(b)$ is called a local maximum, if it is the maximum of the restriction $f\colon (-r+b, r+b) \to \mathbb{R}$ for some $r > 0$. "Maxima" is the plural of maximum.

Theorem 5.6.1 (Rolle) *If f is a differentiable function on an open set containing the closed interval $[a, b]$ with $a < b$. If $f(a) = f(b)$, then there exists $c \in (a, b)$ such that $f'(c) = 0$.*

Proof. Let $\max\{f(x) \colon a \leq x \leq b\}$ denotes the largest value of $f \colon [a, b] \to \mathbb{R}$. Such a value must exist as a real number, by theorems concerning the continuous functions and compact sets. If the maximum occurs at a point in the open interval (a, b) then we are done by Proposition 5.6.1. If not, then either the minimum occurs in the open interval (a, b) or the function f is constant on the interval $[a, b]$. (These two cases are not mutually exclusive). In any case the conclusion follows. □

This theorem generalizes to:

Theorem 5.6.2 (Mean-Value Theorem) *If $y = f(x)$ defines a differentiable function on an open set A containing the closed interval $[a, b]$ with $a < b$. Then there exists $c \in (a, b)$ such that*

$$\frac{f(b) - f(a)}{b - a} = f'(c).$$

Proof. Let

$$g(x) := f(x) - \frac{f(b) - f(a)}{b - a}(x - a) - f(a).$$

Then g is a differentiable function on A, and $g(a) = 0 = g(b)$. Thus Rolle's theorem above implies that there is $c \in (a, b)$ with $g'(c) = 0$. This implies the desired conclusion. □

This simple looking theorem has turned out to be foundational to Calculus. One thing is immediate: the mean value theorem explains why the function is increasing where its derivative takes positive values. Of course, we shall see that there are more merits, later in this book.

5.7 Graphing $y = f(x)$

Graphing (roughly) the given function is a meaningful exercise, because it shows the behavior of the function geometrically/pictorially. The method is based upon:

(1) The mean value theorem, which decides increasing and decreasing properties.

(2) The convexity by the second derivative.

(3) Behavior analysis of the function at the singular points, including the asymptotic behavior.

5.7.1 Increase/decrease

Definition 5.7.1 Let f be a function defined on an interval I in the real line $\mathbb{R}$. f is said to be *increasing*, if $f(a) \leq f(b)$ for any $a, b \in I$ with $a < b$. In contrast, f is called *decreasing* if $f(a) \geq f(b)$ when $a < b$ in I.

Sometimes, we say that f is *strictly increasing* if $f(a) < f(b)$ whenever $a < b$ in I. Of course the concept of strictly decreasing function is defined accordingly.

Theorem 5.7.1 (First derivative test) *Let $a, b \in \mathbb{R}$ be with $a < b$. If $f\colon (a,b) \to \mathbb{R}$ is a smooth function (of class C^1), and if $f'(x) > 0$ for every $x \in (a,b)$, then f is strictly increasing on the interval (a,b). If $f' < 0$ instead, then f is decreasing.*

This follows directly by the mean value theorem. Thus the first step of graphing $y = f(x)$ is to find

(a) The domain of definition of f.

(b) The points at which f is not differentiable.

(c) The points at which $f' = 0$.

These will help us to divide the domain into several intervals and points. Moreover, on each interval, one can decide whether the graph is increasing or decreasing. Then a rough form of graph may be sketched.

Example 5.7.1 Describe roughly the graph of $y = x^3 - 2x + 1$.

We leave it as an exercise.

5.7.2 Convexity

As one sees in the preceding example, one feels that the information on f discussed in the preceding section may be far from being sufficient. Then the following can help.

Definition 5.7.2 A function $y = f(x)$ defined on the closed interval $[a, b]$ with $a < b$ is called *convex* if, for any $r, s \in [a, b]$, the inequality

$$f((1-t)r + ts) \leq (1-r)f(r) + tf(s), \quad forallt \in [0,1]$$

holds.

If $y = -f(x)$ is convex, then f is called *concave.*

Theorem 5.7.2 (The second derivative test) *Let $f\colon (a, b) \to \mathbb{R}$ be a smooth function of class C^2. Then, $f''(x) \geq 0$ for all $x \in (a, b)$ if, and only if, f is convex. On the other hand, $f'' \leq 0$ intead, if and only if the function is concave.*

We choose to omit the proof. Instead we discuss the effect of this to the graphing.

In addition to the discussion of the preceding section, one now solves for x in the equation $f''(x) = 0$. These points will subdivide the intervals considered there. Investigating the sign of $f(x)$ in those subdivided intervals, one can tell the convexity/concavity of the graph. This will help adding more information to the rough sketch of the graph.

5.7.3 Singular critical points

For the rational function such as

$$f(x) = \frac{x^2 - 1}{x + 2}$$

the domain of definition is the union of two unbounded intervals

$$I_1 := \{x\colon x < -2\} \text{ and } I_2 = \{x\colon x > -2\}$$

and the function f is undefined at $x = -2$. Analyzing the behavior of $f(x)$ near such a point may help sketching the graph of f. For instance, if one considers the behavior of $f(x)$ as x approaches -2 from the range $x < -2$, then $f(x)$ will approach $-\infty$. If x approaches -2 from the range $x > -2$, then $f(x)$ approaches $+\infty$. Thus the behavior of $y = f(x)$ is better understood.[5]

[5]This procedure reminds the reader of the concepts called *left-limit* and *right-limit* which are defined as follows:

$$\lim_{x \uparrow a} f(x) = L$$

if, for any $\epsilon > 0$, there exists $\delta > 0$ such that $|f(x) - L| < \epsilon$ for every x with $-\delta < x - a < 0$. Then $\lim_{x\downarrow a} f(x) = L$ is defined by: the same sentence except the final condition is with $0 < x - a < \delta$. If L were $-\infty$ then the condition $|f(x) - L| < \epsilon$ would be replaced by $f(x) < -\epsilon$ instead. In such a way, $\lim_{x\downarrow a} f(x) = L$ (or $\pm\infty$), $\lim_{x\to\pm\infty} f(x)$ etc., and various limits can be considered.

5.8 L'Hôpital's rule

Theorem 5.8.1 (L'Hospital) *If $f(x)$ is smooth of class C^1 in an open interval $(-r+a, a+r)$ for some $r > 0$ satisfying the following three properties:*

(1) $g(x)$ *is not identically zero.*

(2) $f(a) = 0 = g(a)$.

(3) $\lim_{x \to a} \frac{f'(x)}{g'(x)}$ *exists.*

then

$$\lim_{x \to a} \frac{f(x)}{g(x)} = \lim_{x \to a} \frac{f'(x)}{g'(x)}.$$

Proof. If $x \in (-r + a, a + r)$, then there exists $c_1(x), c_2(x)$ between x and a such that

$$f(x) = f(x) - f(a) = f'(c_1(x))(x - a)$$

and

$$g(x) = g(x) - g(a) = g'(c_2(x))(x - a)$$

by the mean value theorem. Notice that both $c_1(x)$ and $c_2(x)$ converge to a, as x approaches a. Therefore

$$\lim_{x \to a} \frac{f(x)}{g(x)} = \lim_{x \to a} \frac{f'(c_1(x))}{g'(c_2(x))} = \lim_{x \to a} \frac{f'(x)}{g'(x)}.$$

if the last limit exists. □

This theorem was generalized to various cases: including the cases that the $\lim f = \lim g = \pm\infty$ and even when the limit is one-sided. But we do not include them here.[6]

---

[6] Guillaume de L'Hospital was a nobleman in French royal court. As French pronunciation habit goes, the middle "s" often stays mute. In modern day spelling, "os" is sometimes replaced by "ô".

5.9 Integration

5.9.1 Indefinite integral

The concept of antiderivative of a function has been discussed earlier. If f admits a function F on the same domain of definition satisfying

$$\frac{d}{dx}F(x) = f(x) \tag{5.3}$$

then we called F the *antiderivative* (or, *primitive*, or *indefinite integral* of f. But this "definition" is not logically sound, since it is unclear whether the function F in (5.3) should be decided precisely enough to be given such a name/terminology.

At the danger of repeating the triviality, we point out that if $F(x)$ satisfies (5.3), so does $F(x) + C$ for any constant C. But would there not be other primitives to the given function f? This question has to be answered.

If F and G satisfy the equation (5.3) for $f(x)$, then we have

$$\frac{d}{dx}(F(x) - G(x)) = 0.$$

We knew that the derivative of any constant function is zero. But if its derivative is zero, then why must the function be constnat?

Proposition 5.9.1 *If $a < b$ and if a differentiable function $f\colon (a,b) \to \mathbb{R}$ satisfies the equation $f'(x) = 0$ for every $x \in (a,b)$, then $f(x)$ is a constant function.*

Proof. Let $p \in (a,b)$. Then the mean value theorem implies that for any $x \in (a,b)$, there exists $c \in (x,p) \cup (p,x)$ such that

$$f(x) - f(p) = f'(c)(x - p).$$

Notice that $c \in (a,b)$. (If $x < p$ then $(p,x) = \varnothing$ and so $x < c < p$. Then $a < x < c < p < b$. The other case also implies $a < c < p$.) Hence by hypothesis $f'(c) = 0$. So $f(x) = f(p)$ for any $x \in (a,b)$. This implies that f is a constant function on the interval (a,b). $\square$

The discussions so far imply that if a function $f\colon (a,b) \to \mathbb{R}$ has a primitive $F(x)$ on the interval (a,b), then $F(x)$ is uniquely determined except by addition of a constant. Thus the definition of the antiderivative (or, primitive) is logically sound.

The primitive (i. e., the antiderivative) of a function f (on an open interval) is also called the *indefinite integral*, the name of which is justified at the end of this section.

5.9.2 Volume by slices

Traditionally, the concept of volume/area by slices has been considered and studied approximately 2,000 years before the understanding of the concept of antiderivative.[7]

The solution was difficult: First, one divides the upper hemisphere by the horizontal slices of thickness $\frac{r}{n}$. When the integer n is sufficiently large, the volumes of slices are almost

$$s_n := \pi\left(r^2 - \left(\frac{kr}{n}\right)^2\right)\frac{r}{n}$$

for $k = 1, \ldots, n$. Then take the total sum of these to obtain

$$W_n = \sum_{k=1}^{n} s_n = \frac{\pi r^3}{n^3}\sum_{k=1}^{n}(n^2 - k^2) = \pi r^3\left(1 - \frac{n(n+1)(2n-1)}{6n^3}\right).$$

Since this is (approximately) the half of the volume of the sphere (of radius r), the volume V_r has to be(!)

$$V_r = 2\lim_{n\to\infty} W_n = \frac{4}{3}\pi r^3.$$

The legend tells that Archimedes was so proud of this and so this was engraved in the tombstone of his! [8] [9]

---

[7]The volume of the sphere of radius r being $\frac{4}{3}\pi r^3$ was known to Archimedes of Syracuse (part of modern Sicily, Italia). History tells us that it was approximately B. C. 250 and modern Calculus was known after a couple of millenia from then.

[8]What a great discovery! But this is only the sphere case, even though a precise proof is still rquired in the light of modern mathematics. Then, what about more complicated general solids?

[9]$\lim_{n\to\infty} A_n = L$ should be understood by the definition similar to the case introduced earlier: for any $\epsilon > 0$, there exists $N > 0$ such that $|A_n - L| < \epsilon$ for every $n > N$. This definition is good, if L is finite. If one want to define $\lim_{n\to\infty} A_n = +\infty$, for instance, then one uses: For any $B > 0$ there exists N such that $A_n > B$ for any $n > N$. Likewise, one can define all cases of various limits using such types of logical conditions.

5.9.3 Definite integral in the sense of Riemann

The problem of computing the volume of the given sphere was understood more systematically with the development of Calculus. The typical problem, in modern mathematical language, is to find the area, say A, between the vertical lines $x = a$ and $x = b$ (for the time being let us say $a < b$) as well as between the x-axis and the graph of $y = f(x)$. Again, for the moment, assume that $f(x) > 0$ on the closed interval $[a, b]$.[10]

The method was taken as follows: First consider a positive integer N and points on the interval $[a, b]$ ordered as

$$a = p_0 < p_1 < \cdots < p_N = b.$$

Such arrangements of points are called "partitions" and denote by P_N. and points $p_0^*, \ldots, p_{N-1}^*$ with $p_i < p_i^* < p_{i+1}$ for every i. (Denote this arrangement of points by P_N^*.) Let

$$S_a^b(N, P_N, P_N^*, f) := \sum_{i=0}^{N-1} f(p_i^*) \cdot (p_{i+1} - p_i).$$

which depends upon the integer N, the partition P_N, the "marking" P_N^* and the function f. This is traditionally called the *Riemann sums* of f on $[a.b]$ given by the partition P_N and the set P_N^* (with no name in particular, but let us call them "markings" for the moment). One introduces the concept of *mesh* $\mu(P_N)$ of each partition as the maximum of $p_{i+1} - p_i$ as $i \in \{0, \ldots, N-1\}$. Then for any sequence of such partitions, we expect to obtain the value for

$$\lim_{\mu(P_N)\to 0} S_a^b(N, P_N, P_N^*, f), \tag{5.4}$$

and expect further that this limit value be uniquely determined regardless of choices for P_N and P_N^*. Notice that this construction does not divide the closed interval $[a, b]$ evenly, but only requires that the mesh shrinks to 0, in which case N tends to ∞ naturally. This gives not much difference here, but this shows its merit when one consider the case when the theory would like to include discontinuous f also. [This is also called the "Riemann integral" as in the case of evenly dividing partitions.]

[10] For such construction, f need not be continuous.

Definition 5.9.1 If the limit value (5.4) exists as a unique finite value B, then we call B the *definite integral* of f on the closed interval $[a, b]$. In such a case, $f\colon [a, b] \to \mathbb{R}$ is *Riemann integrable*, of course.

Of course, this definition is somewhat "irresponsible" because checking whether f is Riemann integrable is not a reasonable task when the function $f\colon [a, b] \to \mathbb{R}$ is given. So one has to streamline the definition. For this purpose, let us limit ourselves to the case when the function f admits a constant C satisfying

$$|f(x)| < C, \text{ for any } x \in [a, b].$$

Such f is said to be *bounded.* [Such a limitation is reasonable, as the nature of Riemann sum requires it in order for its value to be finite.]

Lemma 5.9.1 *If a, b are real numbers with $a < b$, then any bounded function $f\colon [a, b] \to \mathbb{R}$ has supremum and infimum, defined by*

$$\sup_{x\in[a,b]} f(x) := \mathrm{lub}\{f(x)\colon x \in [a, b]\}$$
$$\inf_{x\in[a,b]} f(x) := \mathrm{glb}\{f(x)\colon x \in [a, b]\}$$

where lub *represents the least upper bound, and* glb *the greatest lower bound, respectively.*

This lemma follows by the least upper bound property (an Archimedean axiom). Then define the *upper sums* and the *lower sums* by

$$U_a^b(N, P_N, f) := \sum_{n=1}^{N} \Big(\sup_{x\in[p_{n-1},p_n]} f(x)\Big)(p_n - p_{n-1})$$
$$L_a^b(N, P_N, f) := \sum_{n=1}^{N} \Big(\inf_{x\in[p_{n-1},p_n]} f(x)\Big)(p_n - p_{n-1}),$$

respectively. Then $L_a^b(N, P_N, f) \le S_a^b(N, P_N, P_N^*, f) \le U_a^b(N, P_N, f)$, for all partial sums $S_a^b(N, P_N, P_N^*, f)$.

Lemma 5.9.2 *The definite integral of of f on the closed interval $[a, b]$ exists and is equal to L, if*

$$\lim_{\mu(P_N)\to 0} L_a^b(N, P_N, f) = \lim_{\mu(P_N)\to 0} U_a^b(N, P_N, f) = L.$$

The proof is obvious, but this proposition helps making the definition of definite integral more efficiently as follows:

Definition 5.9.2 (Definite integral) Let $f\colon [a,b] \to \mathbb{R}$ be a function. If

$$\lim_{\mu(P_N)\to 0} L_a^b(N, P_N, f) = \lim_{\mu(P_N)\to 0} U_a^b(N, P_N, f) = B$$

for some real number B, then we say that B is the *definite integral* of f on the closed interval $[a,b]$. In such a case, the function $f\colon [a,b] \to \mathbb{R}$ is called *Riemann integrable*, and B is denoted by

$$\int_a^b f(x)dx.$$

Now the reader realizes that the following questions are not easy to answer but should be important in laying the foundation of Calculus:

- Which functions are Riemann integrable?
- Which functions admit antiderivatives?
- What are the efficient ways of handling the definite integrals and the antiderivatives?

5.10 Fundamental theorem of Calculus

Some answers to the questions posed at the end of preceding section are as follows:

Theorem 5.10.1 *Every continuous function $f\colon [a,b] \to \mathbb{R}$ is Riemann integrable.*

For this proof, we need one fact: for the continuous function such as f above restricted upon a closed and bounded interval $[a,b]$, for any $\epsilon > 0$, there exists $\delta > 0$ such that $|f(x) - f(y)| < \epsilon$ for any $x, y \in [a,b]$ with $|x - y| < \delta$. This property, called the *uniform continuity*, is more than the continuity. But we shall not prove this fact here. Instead we use it to present:

Proof of Theorem 5.10.1: For any partition $P : a = p_0 < p_1 < \ldots < p_n = b$ of $[a,b]$ let the upper and lower Riemann sums be, respectively,

$$U_a^b(n, P, f) = \sum_{k=1}^{n} f(x_{M,k})(p_k - p_{k-1})$$

and

$$L_a^b(n, P, f) = \sum_{k=1}^{n} f(x_{m,k})(p_k - p_{k-1}),$$

where $x_{m,k}, x_{M,k} \in [p_{k-1}, p_k]$ satisfy

$$f(x_{m,k}) \le f(x) \le f(x_{M,k}), \ \forall x \in [p_{k-1}, p_k].$$

Then by the uniform continuity discussed above we see that,

$$f(x_{M,k}) - f(x_{m,k}) = |f(x_{M,k}) - f(x_{m,k})| < \epsilon \tag{5.5}$$

whenever mesh$(P) < \delta$. Then, (5.5) implies that

$$\begin{aligned}
0 < U_a^b(N, P, f) - L_a^b(N, P, f) \\
&= \sum_{k=1}^{n} f(x_{M,k})(p_k - p_{k-1}) - \sum_{k=1}^{n} f(x_{m,k})(p_k - p_{k-1}) \\
&= \sum_{k=1}^{n} \big(f(x_{M,k}) - f(x_{m,k})\big)(p_k - p_{k-1}) \\
&< \sum_{k=1}^{n} \epsilon(p_k - p_{k-1}) \\
&= \epsilon(b-a),
\end{aligned}$$

whenever $\mu(P) :=$ mesh $(P) < \delta$. Therefore the desired conclusion follows, as soon as the limit $\lim_{\mu(P)\to 0} L_a^b(P, f)$ exists (and is finite). We choose to omit the proof of this last, which is an interesting and challenging exercise for the reader. (Hint: Use the least upper bound property.) □

Remark 5.10.1 As the student has learned from experiences, many discontinuous functions are Riemann integrable. There is much more general a result, by H. Lebesgue, on which functions are Riemann integrable. But we refrain from any further discussions.

Theorem 5.10.2 (Fundamental Theorem of Calculus, I) *For a continuous function $f\colon [a,b] \to \mathbb{R}$, $F(x)$ defined by $F(x) := \int_a^x f(x)dx$ for each $x \in [a,b]$ is a (real-valued) function defined on $[a,b]$. Moreover, F is a differentiable function on (a,b) and*

$$\frac{d}{dx}\int_a^x f(x)dx = f(x).$$

Namely, $F(x)$ given as such be the Riemann (definite) integral is the antiderivative of f (up to a constant added) on the open interval (a,b).

Proof. Fix $x \in (a, b)$. Let $\epsilon > 0$. Then

$$\begin{aligned}
\Big|\frac{1}{h}\Big(\int_a^{x+h} f(t)dt - \int_a^{x} f(t)dt\Big) - f(x)\Big| &= \Big|\frac{1}{h}\int_x^{x+h} f(t)dt - f(x)\Big| \\
&= \frac{1}{|h|}\Big|\int_x^{x+h} (f(t) - f(x))dt\Big| \\
&\le \frac{1}{|h|}\Big|\int_x^{x+h} |f(t) - f(x)|dt\Big|.
\end{aligned}$$

Since f is continuous, in particular at x, there exists $\delta > 0$ such that

$$|f(t) - f(x)| < \epsilon, \text{ for any } t \text{ with } |t - x| < \delta.$$

So, whenever t is between x and $x + h$, $|t - x| < \delta$ if $|h| < \delta$. This yields that

$$\begin{aligned}
&\Big|\frac{1}{h}\Big(\int_a^{x+h} f(t)dt - \int_a^{x} f(t)dt\Big) - f(x)\Big| \\
&\qquad \le \frac{1}{h}\Big|\int_x^{x+h} |f(t) - f(x)|dt\Big| \\
&\qquad < \frac{1}{h}\Big|\int_x^{x+h} \epsilon dt\Big| \\
&\qquad = \epsilon,
\end{aligned}$$

whenever $|h - 0| = |h| < \delta$. This shows that

$$\lim_{h\to 0}\frac{1}{h}\Big(\int_a^{x+h} f(t)dt - \int_a^{x} f(t)dt\Big) = f(x).$$

which is what we wanted to prove. □

Corollary 5.10.3 (Fundamental Theorem of Calculus, II) *If $f : [a, b] \to \mathbb{R}$ is continuous and if F is its antiderivative on the open interval (a, b), then for any $[p, q] \subset (a, b)$ it holds that*

$$\int_p^q f(x)dx = F(q) - F(p).$$

Proof. By the uniqueness of antiderivatives (up to an addition of a constant) and the fundamental theorem of calculus I above, we see that

$$F(x) = \int_p^x f(t)dt + C$$

for some constant C. (In fact, the constant is $F(p)$. Thus

$$\begin{aligned} F(q) - F(p) &= \Big(\int_p^q f(t)dt + C\Big) - \Big(\int_p^p f(t)dt + C\Big) \\ &= \int_p^q f(t)dt, \end{aligned}$$

as desired. □

Remark 5.10.2 The method of finding the exact volume/area by slices was very difficult since one has to find the sum of complicated series, which requires various techniques fitting to each case. Even the sphere case of Arichimedes was not so simple. But this fundamental theorem of calculus provides a way (oftentimes easier) to use the antiderivative. This is already a contribution. But the merit of this fundamental theorem is much more; it opens the view. For instance, the rate of change of volume/are is understood through calculus. Whole theory of classical physics became explained through calculus. Thus the theorem deserves to be called "fundamental."

Here also, we shall use the usual notation

$$F(x)\Big|_a^b = \Big[F(x)\Big]_a^b := F(b) - F(a).$$

Corollary 5.10.4 (Integration by parts and by substitution) *If f, u are differentiable with the appropriate domains and ranges, then*

(1) $\int_a^b f'(x)g(x)dx = f(x)g(x)\Big|_a^b - \int_a^b f(x)g'(x)dx.$

(2) $\int_a^b f(u(t))u'(t)dt = \int_{u(a)}^{u(b)} f(x)dx.$

The proof is left to the reader. The key is that the definite integral is understood via the antiderivatives!

Example 5.10.1 *Compute the volume of the sphere of radius r.*

Remark 5.10.3 The definite integral, developed from the consideration of volume by slices (thanks to Archimedes of Syracuse, 250 B.C., approximately), was the integral traditionally for a couple of millenia. Then by the fundamental theorem of calculus, it is evaluated in many cases by the antiderivative (= the primitive), whenever the limits (=the endpoints of the interval of integral) are given. Thus the name "indefinite integral" is justified.

Chapter 6

Some transcendental functions

6.1 Logarithmic function

6.1.1 Construction

Let $P := \{x \in \mathbb{R} \colon x > 0\}$. Define the function $f \colon P \to \mathbb{R}$ by

$$\ln(x) := \int_1^x \frac{1}{t} dt$$

for every $x \in P$. Since the function $g(x) = \frac{1}{x}$ is continuous, the function $f(x)$ is well-defined for every $x \in P$. Fundamental theorem of calculus implies that $\frac{d}{dx} \ln(x) = \frac{1}{x}$. Also $\ln(1) = 0$ by definition of the definite (Riemann) integral.

Moreover, one observes:

Proposition 6.1.1

$$\ln(ab) = \ln(a) + \ln(b)$$

for any $a, b > 0$.

Proof. By definition

$$\begin{aligned} \ln(ab) &= \int_1^{ab} \frac{dt}{t} \\ &= \int_1^a \frac{dt}{t} + \int_a^{ab} \frac{dt}{t}. \end{aligned}$$

Substituting by $u = at$ (and applying Corollary 5.10.4), one obtains

$$\int_a^{ab} \frac{dt}{t} = \int_1^b \frac{du}{u} = \ln b.$$

Thus the conclusion follows. □

Altogether, we have constructed $\ln x$ by definite integral as the function defined on the set P of positive real numbers, satisfying

(1) $\frac{d}{dx} \ln x = \frac{1}{x}$.

(2) $\ln 1 = 0$.

(3) $\ln ab = \ln a + \ln b$.

6.2 Inverse function

Theorem 6.2.1 (Inverse function theorem) *Let $a < b$ and let $(a,b) = \{x \in \mathbb{R}\colon a < x < b\}$. If $f\colon (a,b) \to \mathbb{R}$ is a smooth function (of class C^k) for some $k \in \{1, 2, \ldots, \infty, \omega\}$ with $f'(x) \neq 0$ for any $x \in (a,b)$, then for the set (called the range of f) $R = f\big((a,b)\big) := \{f(x)\colon x \in (a,b)\}$ there is a unique function $g\colon R \to \mathbb{R}$ satisfying the following:*

(1) $f(g(y)) = y$ and $g(f(x)) = x$ for any $y \in R$ and any $x \in (a,b)$.

(2) g is as smooth as f, i. e., $g \in C^k$ if $f \in C^k$.

Such g is called the *inverse function* of f, and denoted by $g = f^{-1}$. Continuing with the notation of this theorem, write $x = g(y)$ ($\Leftrightarrow y = f(x)$). Then $x = g(f(x))$, and by Chain rule we see that

$$1 = \frac{dg}{dy}\frac{dy}{dx} = g'(y)f'(x).$$

Notice that the main thrust of the inverse function theorem is *on the differentiability of the inverse function*, especially at this juncture.

In particular we have:

Corollary 6.2.2 *With the notation of this section*

$$(f^{-1})'(y) = \frac{1}{f'(f^{-1}(y))}.$$

for every $y \in R$.

Notice that, by continuity, f in Theorem 6.2.1 has its derivative positive everywhere, or negative everywhere. Thus the mean-value theorem implies that f is 1-1, namely $f(x) \neq f(y)$ whenever $x \neq y$. It is easy to see that g is also 1-1.

6.3 Exponential function

6.3.1 Construction

Notice that $(\ln x)' = \frac{1}{x} > 0$ for every $x \in P$. Therefore, $f(x) := \ln x$ is a smooth function of class C^∞ and is monotone increasing. Therefore by the inverse function theorem (Theorem 6.2.1) it has the inverse function, which we denote momentarily by $g(x)$.

The range of g is the domain of definition of f, and so $g(y) > 0$ everywhere. We denote by

$$\exp(y) := g(y).$$

Since $\ln 1 = 0$, we must have $\exp(0) = 1$. Moreover $\ln(ab) = \ln(a) + \ln(b)$ implies that, with $\ln(a) = \alpha, \ln(b) = \beta$,

$$\exp(\alpha + \beta) = \exp(\ln(a) + \ln(b)) = \exp(\ln(ab)) = ab = \exp(\alpha)\exp(\beta).$$

6.3.2 The range of ln and the domain of exp

We observed moment ago that identifying the range of the logarithmic function ('log' function, for short hand) is worth while finding. It is known, by the continuity and the injectivity (namely, the 1-1 property, that the image of ln is an interval[1] (not necessarily bounded). Now we present:

Proposition 6.3.1 *The range of the log function* ln *satisfies*

$$\text{Range}(\ln) := \{\ln x \in \mathbb{R} \colon x > 0\} = \mathbb{R}.$$

Proof. Choose sufficiently large an integer m. Then by Proposition 6.1.1,

$$\ln(2^m) = \ln(2 \cdot 2 \cdots 2) = \ln 2 + \ln 2 + \ldots + \ln 2 = m \ln 2.$$

Since

$$\ln 2 = \int_1^2 \frac{dx}{x} \geq 1 \cdot \frac{1}{2} = \frac{1}{2},$$

we see immediately that (for $\ln 1 = 0$)

$$\ln\left([1,\infty)\right) = \bigcup_{m=1}^{\infty} \ln\left([1, 2^m]\right) \supset \bigcup_{m=1}^{\infty} \left[0, \frac{m}{2}\right] = [0, \infty).$$

[1]This property requires a stricter proof, given usually in general topology.

On the other hand,

$$\ln(2^{-m}) + \ln \frac{1}{2^m} = \ln 1 = 0,$$

and consequently,

$$\ln(2^{-m}) = -\ln 2^m = -m \ln 2 < -\frac{m}{2}.$$

So

$$\ln\big((-\infty, 1]\big) = \bigcup_{m=1}^{\infty} \ln\big([2^{-m}, 0]\big) \supset \bigcup_{m=1}^{\infty} \Big[-\frac{m}{2}, 0\Big] = (-\infty, 0].$$

Altogether, we get the desired conclusion. $\square$

In words, the range of the log function ln is the whole set of real numbers, and so is the domain of definition of the "exponential" function exp. (Of course the range of the exp is the same as the domain of ln which is the set of positive real numbers.

6.3.3 More with the function exp and ln

If we denote by

$$\exp(1) = e,$$

then the meaning of e receives the definition as the number satisfying

$$\int_1^e \frac{dt}{t} = 1.$$

On the other hand, for any positive integer m it follows by the discussion of Section 6.3.1 that

$$\exp m = \exp(1 + 1 + \ldots + 1) = \exp(1) \cdots \exp(1) = e^m.$$

Then by logic, we obtain

$$e^{-m} = \frac{1}{e^m}, \quad e^{\frac{q}{p}} = \sqrt[p]{e^q},$$

for any positive integers m, p, q. Also

$$e^0 = 1, \quad 1 = \exp 0.$$

Therefore the value of the function $\exp(x)$ coincides with e^x for every rational number value of x. Thus the function $\exp(x)$ is the extension of the function e^x defined on the rational numbers. Moreover $\exp(x)$ is a smooth function of class C^∞. This suggests that, by way of the just constructed function $\exp(x)$, we may define

$$e^x := \exp(x), \ \forall x \in \mathbb{R}.$$

What about the derivative then?

Proposition 6.3.2 $\frac{d}{dx}e^x = e^x$.

Proof. By the chain rule and the construction of ln, exp we see that

$$1 = \frac{d}{dx}(x) = \ln\big(\exp(x)\big) = \frac{1}{\exp(x)} \cdot \frac{d}{dx}(\exp(x)).$$

This completes the proof. □

At this point, it may be convenient to use the notation by *Bourbaki*: $\mathbb{Z}$ denotes the set of integers, $\mathbb{Q}$ the set of rational numbers, etc.

For the general exponent, one may define (based upon what we already knew from the algebraic construction of a^x for $x \in \mathbb{Q}$) a^x by

$$a^x = e^{x \ln a}, \ \forall x \in \mathbb{R}.$$

Then for any $a > 0$, a^x defined a smooth function as well (by the Chain rule), and the derivative is given by

$$(a^x)' = a^x \ln a.$$

For the real number c, x^c is defined by $x^c = e^{c \ln x}$, for every real number $x > 0$. This is also a smooth function, and we see that

$$(x^c)' = (e^{c \ln x})' = e^{c \ln x} \cdot \frac{c}{x}, \forall x > 0.$$

This implies $(x^c)' = cx^{c-1}$ for $x > 0$. Notice that this does not follow by the case of $c \in \mathbb{Q}$, unless one goes through very difficult arguments.

Remark 6.3.1 As the logarithm was first discovered and studied by Napier, one may want to develop $\log_a x = y$ from $a^y = x$. But then the immediate difficulty arises beyond the level of rational numbers. Note that the second identity is equivalent to

$$e^{y \ln a} = x \Leftrightarrow y \ln a = \ln x.$$

If $a = 1$ then $\ln x = 0$ and so $x = 1$ and we get to no further steps. If $a > 0$ and $a \neq 1$, then we have

$$\log_a x = y = \frac{\ln x}{\ln a}$$

anther smooth function. Notice that this recovers what we knew about logarithmic and exponential functions.

6.4 Trigonometric functions

6.4.1 The radian measure of the angle

Traditionally, the concept of angle was logically initiated by the Euclidean axiom that defines the right angle, by saying that the right angle is the same anywhere. The angle consists of two half lines (rays) emanating from a single point, traditionally understood as the aperture, or the opening to the rays starting from the zero angle, consisting of two identical rays. The the right angle was understood as the half of the flat angle. From then on, fractional angles (of the flat angle) and their multiple angles were considered. Thus, all the concept of rational angles (based upon that the right angle is the unit), and even negative rational angles, were more or less understood. Then what could be the real angles?

If one of the two rays forming an angle is understood as the positive real axis in the xy-plane (the Descartes plane) then the second ray defines a unique point on the unit circle. Then to this angle, this point (at which the second ray intersects with the unit circle) also cuts out the arc starting from the point $(1, 0)$ and the length of this arc corresponds to the "conventional understanding of the angle". This length pertaining to the arc, therefore to the angle, is called the *radian* measure of the angle. Thus 2π measures the full $360°$, the flat angle is π radian, the right angle $\frac{\pi}{2}$ and so forth. The conversion formula is, therefore,

$$x \text{ radian} = \frac{180}{\pi} x \text{ degree}.$$

From now on, we shall use the radian measure for the angles, unless mentioned otherwise.

6.4.2 Trigonometric functions and their derivatives

Traditional sine and cosine are now defined as functions of a real variable, which takes values in radian. Continuing with the discussion above, the angles are always constructed starting from the positive real axis (the ray emanating from the origin of the xy-plane through the point $(1, 0)$). So the angle then corresponds to its radian measure of the arc on the unit circle starting from $(1, 0)$ and ending at a point, say (a, b).

Notice that the length here is a signed length, which is counted positive when the arc travels on the unit circle counterclockwise and, negative, when clockwise. So the radian measure of angles can range from 0 to $+\infty$ and also to $-\infty$, taking all possible real number values.

If x is the radian measure of the angle starting *from* $(1,0)$ *to* (a,b) on the unit circle, then the sine function $\sin x$ and the cosine function $\cos x$ are defined by

$$\sin x = b, \quad \cos x = a.$$

Since the point $(\cos x, \sin x)$ is always on the unit circle, it follows that

$$(\cos x)^2 + (\sin x)^2 = 1, \ |\sin x| \le 1, \ |\cos x| \le 1$$

regardless of the real value for the variable x. It is simple to check that this definition "extends" (or, coincides with) the traditional sine and cosine. This definition also checks the periodicity of the trigonometric functions. Also, the tangent function is also defined as

$$\tan x = \frac{\sin x}{\cos x}.$$

We shall use $\csc x, \sec x, \cot x$ and the conventional notation $\sin^m x = (\sin x)^m$ for the integer value of m except for the case $m = -1$, reserved for the inverse trigonometric functions which we shall introduce in the next section.

Here we shall establish the calculus of trigonometric functions.

Theorem 6.4.1 (Addition formula) *The identity*

$$\sin(\alpha + \beta) = \sin\alpha\cos\beta + \cos\alpha\sin\beta$$

holds for any $\alpha, \beta \in \mathbb{R}$.

Proof. We first consider the case where α, β are nonnegative and $\alpha + \beta < \frac{\pi}{2}$. Then consider the following picture: First attach two angles α and β so that $\angle CAB = \alpha + \beta$ and $\overline{AM} \perp \overline{BC}$ at the point M. Then $|\Delta ABC| = |\Delta ABM| + |\Delta AMC|$, which implies

$$\frac{1}{2}|\overline{AB}||\overline{AC}|\sin(\alpha+\beta) = \frac{1}{2}|\overline{AB}|||\overline{AM}|\sin\alpha + \frac{1}{2}|\overline{AM}|||\overline{AC}|\sin\beta.$$

Then

$$\sin(\alpha+\beta) = \frac{|\overline{AM}|}{|\overline{AC}|}\sin\alpha + \frac{|\overline{AM}|}{|\overline{AB}|}\sin\beta = \sin\alpha\cos\beta + \cos\alpha\sin\beta.$$

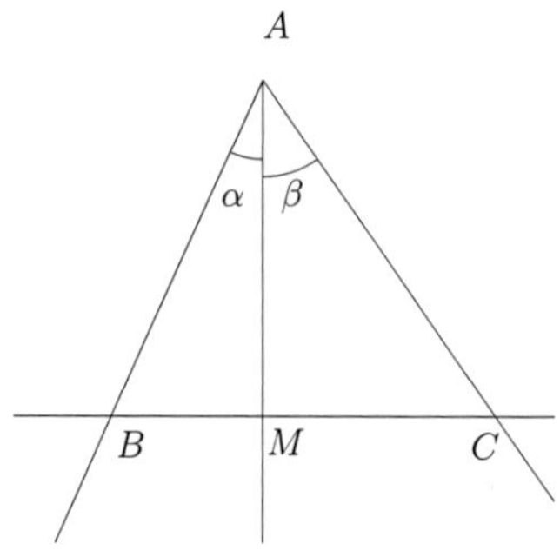

For the case of general angles, one may use the similar arguments and the periodicity to obtain the desired conclusion. □

Let us now see whether $f(x) = \sin x$ is a differentiable function. At any point x, the derivative should be (if exists)

$$f'(x) = \lim_{h\to 0} \frac{1}{h}\big(\sin(x+h) - \sin x\big).$$

By the addition formula (Theorem 6.4.1),

$$\begin{aligned}\sin(x+h) - \sin x &= \sin x \cos h + \cos x \sin h - \sin x \\ &= (\sin x)(\cos h - 1) + \cos x \sin h.\end{aligned}$$

Therefore,

$$f'(x) = \lim_{h\to 0}\Big(\sin x \cdot \frac{\cos h - 1}{h} + \cos x \cdot \frac{\sin h}{h}\Big).$$

Now we show

$$\lim_{h\to 0} \frac{\cos h - 1}{h} = 0 \tag{6.1}$$

and

$$\lim_{h\to 0} \frac{\sin h}{h} = 1, \tag{6.2}$$

which will imply $(\sin x)' = \cos x$. So we prove (6.1) and (6.2). Consider the picture below, which shows (by the area comparison)

$$\frac{1}{2}\cos h \sin h \le \frac{h}{2} \le \frac{1}{2}\tan h.$$

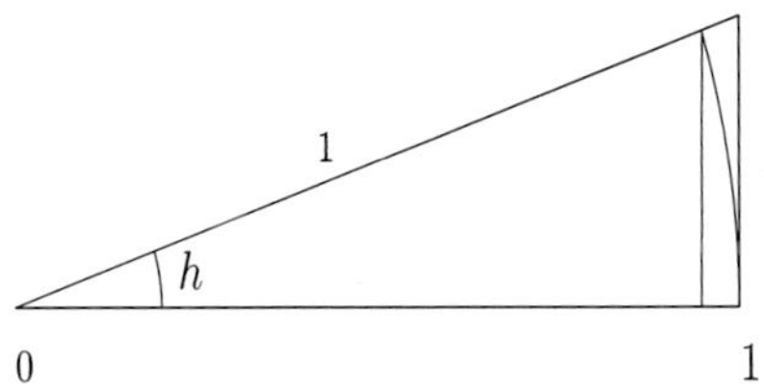

So

$$\cos h \le \frac{h}{\sin h} \le \frac{1}{\cos h}$$

which implies that $\lim_{h\to 0} \frac{\sin h}{h} = 1$. This shows (6.2). Then

$$\begin{aligned} \lim_{h\to 0} \frac{\cos h - 1}{h} &= \lim_{h\to 0} \frac{\cos^2 h - 1}{h(\cos h + 1)} \\ &= \lim_{h\to 0} \sin h \cdot \frac{\sin h}{h} \cdot \frac{-1}{\cos h + 1} \\ &= 0 \cdot 1 \cdot (-1/2) = 0, \end{aligned}$$

which proves (6.1).

Repeating the conclusion, we have checked that

$$(\sin x)' = \cos x.$$

Since $\cos x = \sin(\frac{\pi}{2} - x)$, we immediately obtain

$$(\cos x)' = -\sin x.$$

Then by the rules of differentiation, one gets $(\tan x)' = \sec^2 x$. And one can derive more rules.

Example 6.4.1 Note that $(\sec x)' = \sec x \tan x$. Together with $(\tan x)' = \sec^2 x$ one obtains

$$\sec x = \frac{(\sec x + \tan x)'}{\sec x + \tan x}.$$

From this one finds, using integration by substitution, that

$$\int \sec x \; dx = \ln|\sec x + \tan x| + C.$$

Notice that such formula is not easy to obtain otherwise.[2]

[2]As introduced in Section 1.3, the techniques with half-angle and partial fractions yield a proof to this formula.

6.5 Inverse trigonometric functions

The collection of elementary transcendental functions which most calculus books introduce include the logarithmic and exponential functions, the trigonometric functions, the hyperbolic functions and their inverse functions. The aim of this section is to complete our exposition.[3]

6.5.1 The arcsine function $\sin^{-1} x$

In order to define the inverse function, the original function, say f has to be *injective*, namely, the function f must satisfy t he condition that $f(x) \neq f(y)$ for any x, y in the domain of definition with $x \neq y$. But the sine fıınction is not, because $f(0) = \sin 0 = 0 = \sin \pi = f(\pi)$ for example. So, to discuss the inverse function of $\sin x$, it is customary to restrict the domain to, for instance, the open interval $A = (-\frac{\pi}{2}, \frac{\pi}{2}) := \{x \in \mathbb{R} \colon -\frac{\pi}{2} < x < \frac{\pi}{2}\}$, on which $f(x) = \sin x$ is monotone increasing and, consequently, injective.

Now, it is obvious from the definition of $\sin x$, the range is

$$f\big((-\tfrac{\pi}{2}, \tfrac{\pi}{2})\big) = (-1, 1),$$

an open interval. Therefore the inverse function $f^{-1} \colon (-1, 1) \to (-\frac{\pi}{2}, \frac{\pi}{2})$ is well-defined. By the inverse function theorem (Theorem 6.2.1) this function is as smooth as $f(x)$, and hence, it is of class C^∞. This is the *arcsine function*[4] denoted by

$$\sin^{-1} x = \arcsin x.$$

How about the derivative? Let $y = \sin^{-1} x = f(x)$. Then $x = \sin y = \sin f(x)$. So

$$1 = \frac{d}{dx} \sin f(x) = f'(x) \cos f(x)$$

Since $-\frac{\pi}{2} < f(x) < \frac{\pi}{2}$, $\cos f(x) > 0$ and so

$$\cos f(x) = +\sqrt{1 - \sin^2 f(x)} = \sqrt{1 - x^2}.$$

---

[3]The term *transcendental* emphasizes that they are *beyond* the level of functions that can be constructed from x^r for r rational, by way of four basic operations. The polynomial and rational functions, the multiple roots, and their combinations are called *algebraic* functions. One may find better definition from modern day algebra.

[4]The inverse sine function is also called arcsine, because it assigns, to any $x \in (-1, 1)$, the value $y = \sin^{-1} x$, the angle y that realizes x via $x = \sin y$. The angle value is the length of an arc. Thus the name 'arcsine' is justified.

Therefore

$$\frac{d}{dx}(\sin^{-1} x) = \frac{1}{\sqrt{1-x^2}}, \qquad \int \frac{dx}{\sqrt{1-x^2}} = \arcsin x + C.$$

Note that this formula depends upon the restriction of the domain of definition of $\sin x$. For instance, if the decreasing branch of $\sin x$ were taken initially and if the inverse sine were built based upon such a choice, then naturally the derivative will take negative values.

The arcsine function may take slightly larger a domain of definition that is the closed interval $[-1, 1]$, in which case the range is $[-\frac{\pi}{2}, \frac{\pi}{2}]$. This is a continuous function which is the inverse function of the sine function. But for the purpose of differentiation, the domain is better to be the open interval $(-1, 1)$.

6.5.2 The arctangent

$g(x) := \tan x$ defines a smooth function (of class C^∞) on the open interval $(-\frac{\pi}{2}, \frac{\pi}{2})$ with range $\mathbb{R}$, the whole set of real numbers. g is injective. Therefore, the inverse function (namely, the *arctangent*) is well-defined and is of class C^∞.

The derivative can be also computed, much along the same way.

$$1 = (x)' = \big(\tan(\tan^{-1} x)\big)' = \big(\sec^2(g(x))\big)g'(x).$$

So,

$$\begin{aligned}
(\tan^{-1} x)' &= g'(x) \\
&= \frac{1}{\sec^2(g(x))} \\
&= \frac{1}{1+\tan^2(g(x))} \\
&= \frac{1}{1+\tan^2(\tan^{-1} x)} \\
&= \frac{1}{1+x^2}.
\end{aligned}$$

Other types of inverse trigonometric functions are understood similarly, following the course of exposition given here.

6.6 Hyperbolic functions

The hyperbolic functions are given as follows:

$$\sinh x := \frac{e^x - e^{-x}}{2}$$
$$\cosh x := \frac{e^x + e^{-x}}{2}$$
$$\tanh x = \frac{\sinh x}{\cosh x}.$$

They satisfy

$$\cosh^2 x - \sinh^2 x = 1, \quad 1 - \tanh^2 x = \frac{1}{\cosh^2 x}$$

and

$$(\sinh x)' = \cosh x, \ (\cosh x)' = \sinh x$$

and

$$(\tanh x)' = \frac{\cosh^2 x - \sinh^2 x}{\cosh^2} = \frac{1}{\cosh^2}.$$

Moreover,

$$\tanh^{-1} x = \frac{1}{2} \ln \frac{1+x}{1-x}.$$

(Notice that the range of $\tanh x$ is the open interval $(-1, 1)$.)

Chapter 7

Exams of Semester 2019-01

7.1 Examination 1

Exam 1 (10%) – Math 101
Calculus I, Semester 2019-01, POSTECH
19:00-20:00 March 12 2019

Directions: No calculators are allowed. All problems are of equal weight.

[**1**] Find the integral $\int \sin^3 x \; dx$.

[**2**] Find $\int \frac{x}{(x+2)(x+1)^2} dx$.

[**3**] Find $\int \frac{dx}{5 - 3\cos x}$.

[4] Test the convergence of

$$\sum_{n=1}^{\infty} \frac{5^{3n^2} - 6^n}{5^{3n^2-n} + 7^{n\sqrt{n}}}.$$

[5] Test the convergence of the series:

$$\sum_{n=2}^{\infty} \frac{\cos^n n}{2^n - \sin n^3}.$$

7.2 Midterm Examination

Midterm Exam (30%) – Math 101
Calculus I, Semester 2019-01, POSTECH
18:00-21:00 Monday, April 8, 2019

Please Note: (1) No calculators are allowed.
(2) Grading applies only on what is written on the exam paper.
(3) This exam consists of 10 problems; All problems receive equal weight.
(4) Both $\log x$ and $\ln x$ represent the same *natural logarithmic function.* This examination does *not* use the "common log" function $\log_{10} x$.

[**1**] Determine the values of real number α that makes the series $\displaystyle\sum_{n=1}^{\infty} \frac{(\log n)^{\alpha}}{n}$ converge.

[**2**] Evaluate $\displaystyle\int \frac{1}{(1+x^2)^2} dx$.

[**3**] Compute the radius of convergence and the interval of convergence of the power series

$$f(x) = \sum_{n=0}^{\infty} (-1)^n x^{2n} \quad g(x) = \int_0^x f(y)\, dy,$$

respectively.

[**4**] Find the Taylor series of $f(x) = \log(x + \sqrt{1+x^2})$ about $x = 0$ and also tis radius of convergence.

[**5**] Find the radius of convergence of $\displaystyle\sum_{n=1}^{\infty} a_n x^n$, where $\displaystyle\frac{1}{\sqrt{n}} \leq a_n \leq \sqrt{n}$ for every integer n.

[**6**] Find the function of the form $f(x) = \sum_{n=0}^{\infty} a_n \frac{x^n}{n!}$, which satisfies

$$\frac{d^3}{dx^3} f(x) = -f(x),\ f(0) = 1,\ f'(0) = 0,\ f''(0) = 2.$$

[7] (a) Prove that $\cos x = \sum_{n=0}^{\infty}(-1)^n \frac{x^{2n}}{(2n)!}$.

(b) Find the limit $\lim_{x\to 0} \frac{\cos x^6 - 1 + \frac{1}{2}x^{12} - \frac{x^{24}}{4!}}{x^{36}}$.

[8] Derive the Taylor series of the function $f(x) = \log(2+3x)$ about $x = 0$ and determine the set of points x at which its Taylor series $[T_0 f](x)$ converges to $f(x)$.

[9] Find an approximate value of $\int_0^1 e^{-x^2}dx$ with allowed error not more than 0.01. Justify your answer.

[10] Assume that a_n, for every $n = 0, 1, 2, \cdots$, is an integer between 1 and 10^8. Then, find the radius of convergence of the series $\sum_{n=0}^{\infty}(a_n + 1000\cos^2 n)x^n$.

7.3 Examination 3

Exam 3 (10%) — Math 101
Calculus I, Semester 2019-01, POSTECH
09:00-12:00 Wednesday, 05 June 2019

Please Note: (1) No calculators are allowed.
(2) Grading applies only on what is written on the exam paper.
(3) This exam consists of 10 problems; All problems receive equal weight.
(4) Both $\log x$ and $\ln x$ represent the same *natural logarithmic function.* This examination does *not* use the "common log" function $\log_{10} x$.

[**1**] You are having an iced coffee on a Summer day with the temperature 27° C. According to *Newton's law of cooling*, the temperature $T(t)$ of the coffee at time t satisfies the differential equation

$$\frac{dT}{dt} = -k(T - 27°C),$$

where k is a constant. This coffee was 7° C ten minutes ago, and now it is 17° C. What is the temperature of the coffee after ten minutes from now?

[**2**] Solve the following (logistic) differential equation

$$\frac{dy}{dx} - 2y = -y^2$$

with the initial condition $y(0) = 1$.

[**3**] Solve the following (first order linear) differential equation

$$2\frac{dy}{dx} - y = e^x$$

with the initial condition $y(0) = 2$.

[**4**] Use the ϵ-and-δ definition of limit to prove that

$$\lim_{x \to 2}(-2x + 1) = -3.$$

7.4 Final Examination

Final Exam (50%) — Math 101
Calculus I, Semester 2019-01, POSTECH
09:00-12:00 Wednesday, 05 June 2019

Please Note: (1) No calculators are allowed.
(2) Grading applies only on what is written on the exam paper.
(3) This exam consists of 10 problems; All problems receive equal weight.
(4) Both $\log x$ and $\ln x$ represent the same *natural logarithmic function*. This examination does *not* use the "common log" function $\log_{10} x$.

---

[1] Determine whether $\displaystyle\sum_{n=5}^{\infty} \frac{(n+1)^n}{n^{n+1}(\log n)^2}$ converges.

[2] Determine whether the sequence $A_n = 2+\frac{2}{3}+\frac{2}{5}+\ldots+\frac{2}{2n-1}-\log(2n+1)$ converges as n tends to infinity.

[3] Evaluate $\int x^2 \arcsin x dx$.

[4] Evaluate $\displaystyle\int \frac{x}{(x+1)^2(x+2)}dx$.

[5] Find the radius of convergence of the series $\displaystyle\sum_{n=1}^{\infty} \frac{\sin((n\pi)/2)}{2^n}x^n$.

[6] Find an approximate value of $\displaystyle\int_0^1 \sin(x^2)dx$ with allowed error not more than 0.001. Justify your answer.

[7] Solve for the function y from

$$y' + y = e^{-x+1}, \quad y(0) = e.$$

[8] Let $f : \mathbb{R} \to \mathbb{R}$ be a function defined by

$$f(x) = \begin{cases} x^2 \sin \frac{1}{x} & \text{if } x \neq 0, \\ 0 & \text{if } x = 0. \end{cases}$$

Show that $f(x)$ is differentiable at $x = 0$.

[9] Assume that the solution to the differential equation

$$y' = y^2 + 1$$

with the initial condition $y(0) = 0$ is a power series of type $\sum_{n=0}^{\infty} a_n x^n$ that is convergent for every x with $|x| < r$ for some positive number r. Find the first two nonzero terms of the Taylor series of the solution function centered at 0.

[10] The functions f, g and h are defined on the interval $[-1, 1]$ satisfying $f(x) \leq g(x) \leq h(x)$ for every $x \in [-1, 1]$. Assume also that

$$\lim_{x \to 0} f(x) = 3 = \lim_{x \to 0} h(x).$$

Show, using ϵ-and-δ method (i. e., the definition of limit), that $\lim_{x \to 0} g(x) = 3$.

Part II: Multidimensional Calculus

Course Plan for Semester 2019-02

Weeks 1-2a: Basic geometry [Chapter 8]

Weeks 2b-4: Differentiability, Derivative, Directional derivative, Partial derivative, Gradient, Chain rule [Chapter 9]

Exam 1 (on contents of Week 1 – Week 4).

Weeks 5-6: Taylor expansion and Min-Max principle for $y = f(x_1, \ldots, x_n)$ with the 2nd derivative test, Lagrange multiplier method. [Chapter 10]

Week 7: Applications of multidimensional differential calculus [Ch 11]

Week 8: Midterm examination (on contents of Week 1 – Week 7)

Weeks 9-10: Integration (double and triple integrals, coordinate changes) [Lax-Terrell Ch 6, p205-278] & [KTK's Lecture notes Ch 12]

Weeks 11-13a: Line and surface integrals (including Work, Circulation, Flux) [Lax-Terrell Ch 7; p.279-332.] & [KTK Ch 12.4.5-13],

Exam 3 (in the 12th week on contents of Week 9 – Week 12a)

Weeks 13b-15: Integral theorems [Lax-Terrell Ch 8; p333-386] & [KTK Ch 14]

Week 16: Final exam (on the whole semester course).

Chapter 8

Some geometry of $\mathbb{R}^n$

8.1 Dot product; length and angle

It is a usual practice in Mathematics of modern days that the notation $\mathbb{R}$ stands for the set of real numbers, and $\mathbb{R}^n$ for a positive integer n represents the set

$$\mathbb{R}^n := \{(a_1, \ldots, a_n) \colon a_i \in \mathbb{R}, \forall i = 1, \ldots, n\},$$

of ordered n-tuples of real numbers. We may call each element of $\mathbb{R}^n$ a *vector*, and we may use the notation such as

$$\vec{a} := (a_1, \ldots, a_n).$$

In Physics, this represents the terminal point of the vector $\vec{a}$, whereas the initial point is the origin. [Therefore, all vectors in $\mathbb{R}^n$ are regarded to have their initial point at the origin $\vec{0} := (0, \ldots, 0)$.]

Of course, it is easy to observe that there are operations for $r \in \mathbb{R}$, $\vec{a} := (a_1, \ldots, a_n) \in \mathbb{R}^n$ and $\vec{b} := (b_1, \ldots, b_n) \in \mathbb{R}^n$, inherited from the four basic operations for the real numbers, which are:

(1) $r\vec{a} = r(a_1, \ldots, a_n) = (ra_1, \ldots, ra_n)$.

(2) $\vec{a} + \vec{b} = (a_1 + b_1, \ldots, a_n + b_n)$.

The first is the same as the *scalar multiplication* of a vector $\vec{a}$ by the scalar r, whereas the second the *addition* of vectors.

In $\mathbb{R}^2$, particularly, the *norm* of the vector $\vec{a} := (a_1, a_2)$ is the concept induced from the length of the vector with the endpoint at (a_1, a_2) and initial point at $(0,0)$. Thus the Pythagorean theorem tells us that the norm of $\vec{a}$ is (should be?) defined to be

$$\|\vec{a}\| = \sqrt{a_1^2 + a_2^2},$$

and it is agreed to be natural.

Another operation is the *dot product.* It is defined to be

$$\vec{a} \cdot \vec{b} = (a_1, \ldots, a_n) \cdot (b_1, \ldots, b_n) = a_1 b_1 + \ldots + a_n b_n.$$

Therefore, $\|\vec{a}\| = \sqrt{\vec{a} \cdot \vec{a}}$. In case of $\mathbb{R}^2$, consider two distinct nonzero vectors $\vec{a}$ and $\vec{b}$. What is the significance of this "dot product" then?

In the intuitive way of Physics, the vector with the initial point at (a_1, a_2) and the terminal point at (b_1, b_2) is a parallel translate of the vector $\vec{c} := (b_1 - a_1, b_2 - a_2)$. Together with $\vec{a}$ and $\vec{b}$ these 3 vectors form a triangle.

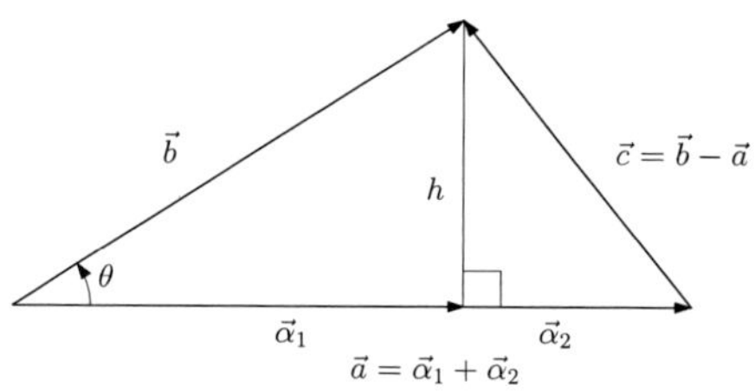

Then, if we denote by θ the angle required to turn $\vec{a}$ to the direction of $\vec{b}$, we see that

$$\|\vec{\alpha}_1\| = \|\vec{b}\| \cos\theta, \text{ and } h = \|\vec{b}\| \sin\theta.$$

Thus, Pythagorean theorem implies

$$\begin{aligned} \|\vec{b} - \vec{a}\|^2 &= \|\vec{c}\|^2 \\ &= \|\vec{\alpha}_2\|^2 + h^2 \\ &= (\|\vec{a}\| - \|\vec{b}\| \cos\theta)^2 + (\|\vec{b}\| \sin\theta)^2 \end{aligned} \tag{8.1}$$

Simplifying the terms (of the first and the last expressions), one obtains

$$\vec{a} \cdot \vec{b} = \|\vec{a}\| \|\vec{b}\| \cos\theta.$$

Remark 8.1.1 This is the result in $\mathbb{R}^2$, but it continues to hold in all dimensions, since a triangle in $\mathbb{R}^n$ for any $n \geq 2$ is in the plane which is the extension of the triangle and hence in a copy of $\mathbb{R}^2$. Checking the detail (if any) is left for the reader.

We might point out in passing that the second expression $||\vec{c}||^2$ is equal to the last expression of (8.1), of course. Simplifying the last expression then yields what is known as the *law of cosines*:

$$||\vec{c}||^2 = ||\vec{a}||^2 + ||\vec{b}||^2 - 2||\vec{a}||\,||\vec{b}||\cos\theta.$$

We may also point out that there is not much distinction between a point in $\mathbb{R}^n$ and a vector in $\mathbb{R}^n$. (Although, to a purist, an element of $\mathbb{R}^n$ is *not* the point itself, but only a *representation* of the point. But we shall not discuss this any further.)

8.2 Lines and (hyper-)planes

Let us go back to middle-school mathematics.[1] In the xy-plane, a *straight line* is the set of points (x, y) consisting of real values of x and y satisfying the relation

$$ax + by = c$$

for some constants a, b and c where at least one of a, b is nonzero.

In Euclidean geometry, one of the ways that describe a line (i. e., a straight line) is as follows:

(†) A line is determined by a *point* (that it passes though) and another *line* that is perpendicular (orthogonal).

This can be explained via the concept of "dot product" which we shall explain in the next section.

8.2.1 As sets

Now, the following definition may be accepted naturally.

Definition 8.2.1 Two vectors $\vec{a}, \vec{b} \in \mathbb{R}^n$ are said to be *orthogonal* (or, equivalently, *perpendicular*) to each other, if $\vec{a} \cdot \vec{b} = 0$.

Line

The description (†) can now receive the description by coordinate system such as:

(‡) The *straight line*, say ℓ, in $\mathbb{R}^2$, passing through a point $\vec{p} := (p_1, p_2)$ orthogonal to the vector $\vec{n} = (n_1, n_2)$ is represented by

$$\ell = \{(x, y) \in \mathbb{R}^2 \colon \vec{n} \cdot \big((x, y) - (p_1, p_2)\big) = 0\}.$$

Note that

$$\begin{aligned} &\vec{n} \cdot \big((x, y) - (p_1, p_2)\big) = 0 \\ &\Leftrightarrow n_1(x - p_1) + n_2(y - p_2) = 0 \\ &\Leftrightarrow n_1 x + n_2 y = n_1 p_1 + n_2 p_2. \end{aligned}$$

[1]But, seriously, this part of discussion is not simple, but actually deep. It was one of the keys for Euclid to formulate geometry which bears his name in history!

This shows that the general expression $ax + by = c$ of line mentioned in the beginning part of this chapter is indeed fitting to a traditional description from Euclidean geometry.

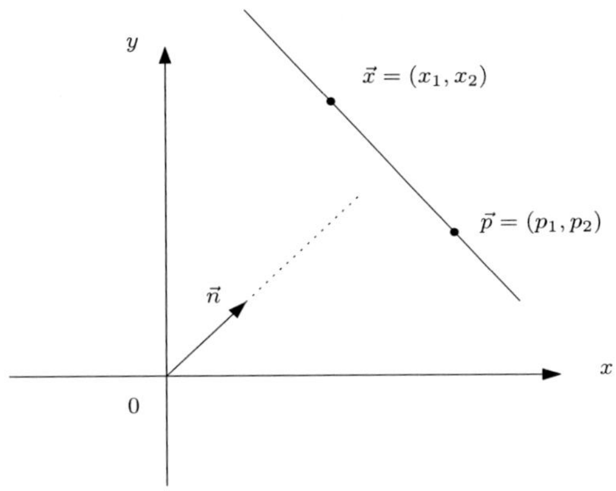

Figure 8.1: Straight line in $\mathbb{R}^2$

Planes in $\mathbb{R}^3$

On the other hand, the expression

$$\vec{n} \cdot (\vec{x} - \vec{p}) = 0,$$

for $\vec{x} = (x_q, x_2, x_3)$, where $\vec{0} \neq \vec{n}, \vec{p} = (p_1, p_2, p_3) \in \mathbb{R}^3$, represents the object in $\mathbb{R}^3$ consisting of vectors with the properties that the "location vector" with its initial point at $\vec{p}$ and final point at $\vec{x}$ perpendicular to (the line generated by) the vector $\vec{n}$—which is a flat *plane*. Also, it receives the general expression

$$n_1(x_1 - p_1) + n_2(x_2 - p_2) + n_3(x_3 - p_3) = 0$$

or equivalently

$$a_1x_1 + a_2x_2 + a_3x_3 = c.$$

Hyperplanes in $\mathbb{R}^n$

Generally in $\mathbb{R}^n$, the expression

$$\vec{n} \cdot (\vec{x} - \vec{p}) = 0$$

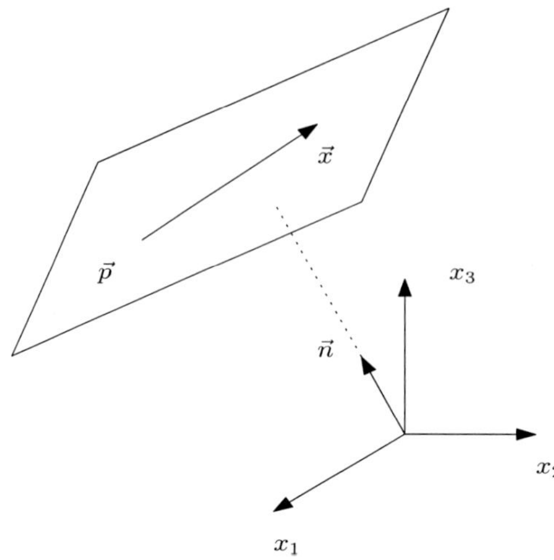

Figure 8.2: Plane in $\mathbb{R}^3$

defines the image of $\mathbb{R}^{n-1} \times \{0\}$ by a translation and rotation (i. e., a rigid motion of $\mathbb{R}^n$)—which is called a *hyperplane.* [2]

8.2.2 By parametrization

Lines in $\mathbb{R}^3$ (also in $\mathbb{R}^n$)

Another way of determining a straight line, say ℓ, is via a point, say $\vec{p}$, which it passes through and a vector, say $\vec{v}$, that is parallel to the line. This means that $\vec{x}$ is on ℓ if, and only if, $\vec{x} - \vec{p} = t\vec{v}$ for some $t \in \mathbb{R}$. Therefore,

$$\ell = \{\vec{x} \in \mathbb{R}^3 \colon \vec{x} = \vec{p} + t\vec{v}, t \in \mathbb{R}\}.$$

Therefore, the parametric equation (for the parameter t taking value in $\mathbb{R}$)

$$\begin{cases} x_1(t) = p_1 + v_1 t \\ x_2(t) = p_2 + v_2 t \\ x_3(t) = p_3 + v_3 t \end{cases}$$

[2]What is a *hyperplane*? It is, of course, an analog of a plane in $\mathbb{R}^3$. As the plane in $\mathbb{R}^3$ is the result of a rotation followed by a parallel translation of the copy of $\mathbb{R}^2$ in $\mathbb{R}^3$, a hyperplane is the result of rigid motion (in $\mathbb{R}^n$) of the copy of $\mathbb{R}^{n-1}$. Since it is not a plane (in the sense that the "dimension" is not 2), the terminology such as *hyperplane* seems to have been accepted.

represents the line passing through $\vec{p} = (p_1, p_2, p_3)$ parallel to the nonzero vector $\vec{v} = (v_1, v_2, v_3)$..

Of course, this idea and exposition is valid in $\mathbb{R}^n$. It is required simply to put the correct coordinate expressions to the vector equation $\vec{x}(t) = \vec{p} + t\vec{v}$. Detailed construction is left to the student.

Planes in $\mathbb{R}^3$

A plane, say P, passing through the origin $\vec{0} = (0, 0, 0)$ contain at least two nonzero vectors $\vec{u}, \vec{v}$ each of which is not a constant multiple of the other, i. e., is not parallel to the other. Then any other point $\vec{x} \in P$ can be written as

$$\vec{x} = s\vec{u} + t\vec{v}$$

where s, t are real numbers. Now, if Π is a plane in $\mathbb{R}^3$ passing through a point $\vec{p}$, then the set

$$P_\Pi := \{\vec{x} - \vec{p} \colon \vec{x} \in \Pi\}$$

is a plane, as it is a parallel translation of Π. P_Π passes through $\vec{0} = \vec{p} - \vec{p}$. Therefore one arrives at the vector equation

$$\vec{x}(s, t) = \vec{p} + s\vec{u} + t\vec{v}.$$

Rewriting it as the standard parametric equation (in two variables s, t) using the components yields, of course, the following parametric equation:

$$\begin{cases} x_1(s, t) = p_1 + u_1 s + v_1 t \\ x_2(s, t) = p_2 + u_2 s + v_2 t \\ x_3(s, t) = p_3 + u_3 s + v_3 t. \end{cases}$$

8.3 Cross product

Another operation for the vectors of $\mathbb{R}^3$ to introduce here is the concept of *cross product* (or, the *vector product*). For the moment let us say that all the vectors are vectors of $\mathbb{R}^3$.

8.3.1 Algebraic definition

Definition 8.3.1 The *cross product* is an operation $(\vec{a}, \vec{b}) \mapsto \vec{a} \times \vec{b}$ between the vectors in $\mathbb{R}^3$, satisfying the following rules:

(1) $(\vec{u} + \vec{v}) \times \vec{w} = \vec{u} \times \vec{w} + \vec{v} \times \vec{w}$.

(2) $(c\vec{v}) \times \vec{u} = c(\vec{v} \times \vec{u})$, for any $c \in \mathbb{R}$.

(3) $\vec{v} \times \vec{u} = -\vec{u} \times \vec{v}$. In particular, $\vec{u} \times \vec{u} = \vec{0}$.

(4) Denote by $\vec{i} = (1, 0, 0), \vec{j} = (0, 1, 0), \vec{k} = (0, 0, 1)$. Then $\vec{i} \times \vec{j} = \vec{k}$, $\vec{j} \times \vec{k} = \vec{i}$, and $\vec{k} \times \vec{i} = \vec{j}$.

Proposition 8.3.1 *The following hold:*

(a) $\vec{v} \cdot (\vec{u} \times \vec{v}) = 0 = \vec{u} \cdot (\vec{u} \times v)$. *Namely,* $\vec{u} \times \vec{v}$ *is perpendicular to both* $\vec{u}$ *and* $\vec{v}$.

(b) $(a_1, a_2, a_3) \times (b_1, b_2, b_3) = (a_2b_3 - a_3b_2, a_3b_1 - a_1b_3, a_1b_2 - a_2b_1)$.

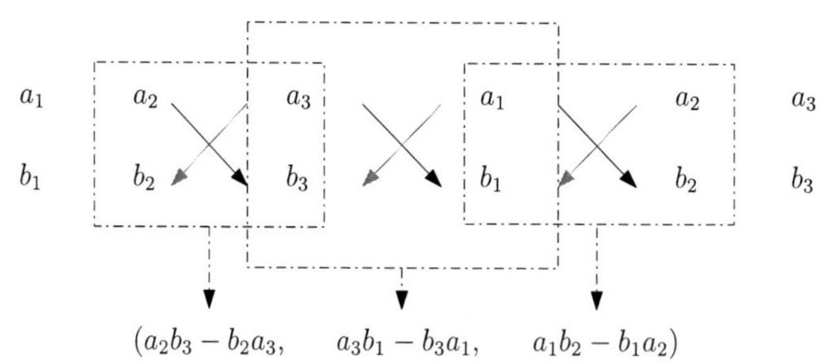

Figure 8.3: How to compute $\vec{a} \times \vec{b}$

8.3.2 Lagrange's identity and consequences

We now list some properties of the cross product that are useful for many purposes:

Proposition 8.3.2 *For vectors in $\mathbb{R}^3$, the following hold:*

(A) $\|\vec{a}\times\vec{b}\|^2 = \|\vec{a}\|^2\|\vec{b}\|^2 - (\vec{a}\cdot\vec{b})^2$.

(B) $(\vec{a}\times\vec{b})\cdot(\vec{c}\times\vec{d}) = (\vec{a}\cdot\vec{c})(\vec{b}\cdot\vec{d}) - (\vec{a}\cdot\vec{d})(\vec{b}\cdot\vec{c})$.

(C) $|\vec{a}\cdot(\vec{b}\times\vec{c})|$ *is equal to the volume of the parallelepiped generated by* $\vec{a},\vec{b}$ *and* $\vec{c}$.

(D) $\vec{a}\times(\vec{b}\times\vec{c}) + \vec{b}\times(\vec{c}\times\vec{a}) + \vec{c}\times(\vec{a}\times\vec{b}) = 0$.

The proofs (especially for (B) and (D)) are initially done by computing directly all terms by components. I know that it may sound stupid but it works! (A) follows by (B), of course. On the other hand, (B) is known as one of *Lagrange's identities.* (C) follows by (A) and some knowledge of old plane geometry that $\|\vec{a}\times\vec{b}\|$ is, due to (A), equal to the area of the parallelogram generated by $\vec{a}$ and $\vec{b}$. [3] By the way, $\vec{a}\cdot\vec{b}\times\vec{c}$ is called the *scalar triple product* and often denoted by $[\vec{a}\vec{b}\vec{c}]$.

Remark 8.3.1 (D) is called the *Jacobi identity*, which may seem odd. But I would emphasize here that the cross product is not associative, namely in general $(\vec{a}\times\vec{b})\times\vec{c}\neq\vec{a}\times(\vec{b}\times\vec{c})$. Then how can one know how to take the cross products of more than 3 vectors? No rules? Yes, you guessed it. This Jacobi identity, even though the product is not associative, can contribute to that! We choose to refrain from providing any further discussions, as we will use only a few properties of the cross product.

[3] For $\vec{a} := a_1\vec{i} + a_2\vec{j}$ and $\vec{b} = b_1\vec{i} + b_2\vec{j}$ $\vec{a}\times\vec{b} = (a_1b_2 - a_2b_1)\vec{k}$. Hence $|\vec{a}\times\vec{b}| = |(a_1,a_2)\cdot(-b_2,b_1)|$. Since $(-b_2,b_1)$ is the rotation of $\vec{b}$ by $+90°$, we obtain that $\|\vec{a}\times\vec{b}\| = \|\vec{a}\|\|\vec{b}\||\sin\theta|$ where θ is the angle between $\vec{a}$ and $\vec{b}$. Hence one can see (not having to appeal to (A) of the proposition above) that $\|\vec{a}\times\vec{b}\|$ represents the area of the parallelogram generated by $\vec{a}$ and $\vec{b}$.

8.3.3 Some applications of cross product

Lines in $\mathbb{R}^3$ as the intersection of two planes

If the two planes are given by vector equations such as

$$(\vec{x} - \vec{p}_1) \cdot \vec{n}_1 = 0 \quad \text{and} \quad (\vec{x} - \vec{p}_2) \cdot \vec{n}_2 = 0$$

respectively, then the intersection is a straight line and is therefore represented by the simultaneous equations. But, one can extract the vector equation of the line, which turns into a parametric equation as well. This is possible because the intersection straight line is in both planes and hence it has to be perpendicular to the vectors $\vec{n}_1$ and $\vec{n}_2$ simultaneously. Hence the line has to be parallel to the vector $\vec{n}_1 \times \vec{n}_2$. Then with one point on this line (i. e., a point, say $\vec{q}$, satisfying both equations of the planes), one obtains

$$\vec{x}(t) = \vec{q} + t(\vec{n}_1 \times \vec{n}_2)$$

as the desired vector equation. Of course in terms of coordinates, this turns into a parametric equation.

Planes in $\mathbb{R}^3$ given by 3 non-collinear points

The points $\vec{p}_1, \vec{p}_2$ and $\vec{p}_3$ are not collinear (i. e., the lines joining any two of these points does not have the remaining 3rd point on it) then $\vec{u} := \vec{p}_1 - \vec{p}_2$ and $\vec{v} := \vec{p}_2 - \vec{p}_3$ are not parallel to each other. Now, $\vec{n} := \vec{u} \times \vec{v}$ will then be a nonzero vector that is perpendicular to the plane passing through the 3 points $\vec{p}_1, \vec{p}_2$ and $\vec{p}_3$. Therefore this yields the equation for $\vec{x}$

$$(\vec{x} - \vec{p}_1) \cdot \vec{n} = 0$$

or equivalently,

$$(\vec{x} - \vec{p}_1) \cdot (\vec{p}_1 - \vec{p}_2) \times (\vec{p}_2 - \vec{p}_3) = 0.$$

8.4 Orthogonal projection and applications

8.4.1 Orthogonal projection

Let $\vec{A}, \vec{B} \in \mathbb{R}^n$ and let $\vec{A} \neq \vec{0}$. Then from the setting of the figure:

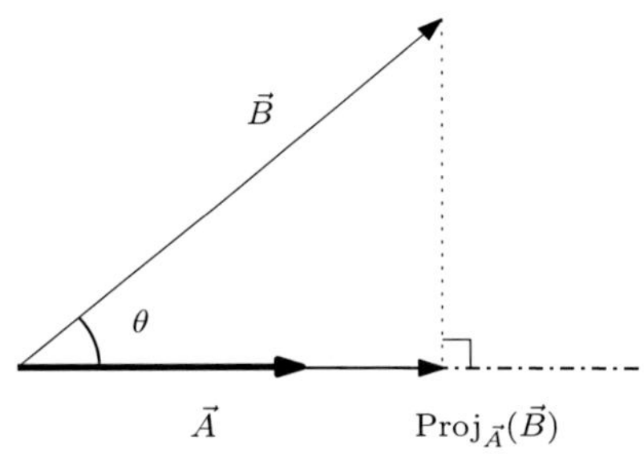

Figure 8.4: Orthogonal projection of $\vec{B}$ onto $\vec{A}$

the orthogonal projection of $\vec{B}$ onto (the line in the direction of) $\vec{A}$ should be $\|\vec{B}\| \cdot \cos\theta \cdot \frac{\vec{A}}{\|\vec{A}\|}$, which can be rewritten such as

$$\begin{aligned}\|\vec{B}\| \cos\theta \frac{\vec{A}}{\|\vec{A}\|} &= \frac{\|\vec{B}\|}{\|\vec{A}\|} \frac{\vec{A}\cdot\vec{B}}{\|\vec{A}\|\|\vec{B}\|}\vec{A} \\ &= \frac{\vec{A}\cdot\vec{B}}{\vec{A}\cdot\vec{A}}\vec{A}.\end{aligned}$$

This leads us to

Definition 8.4.1 Let $\vec{A}$ be a nonzero vector in $\mathbb{R}^n$. Then for any $\vec{v} \in \mathbb{R}^n$, the *orthogonal projection* of $\vec{v}$ onto $\vec{A}$ is defined to be

$$\text{Proj}_{\vec{A}}(\vec{v}) = \frac{\vec{v}\cdot\vec{A}}{\vec{A}\cdot\vec{A}}\vec{A}.$$

8.4.2 Distance between a point and a line, or a plane

A straight line, say ℓ, in $\mathbb{R}^2$ is the collection of all $\vec{x} \in \mathbb{R}^2$ satisfying

$$(\vec{x} - \vec{p}) \cdot \vec{n} = 0$$

where $\vec{n}in\mathbb{R}^2$ is a nonzero vector. Let $\vec{q} \in \mathbb{R}^2$. The distance, say $d(\vec{q}, \ell)$, between q and the line ℓ should be $\|\vec{x}_0 - \vec{q}\|$ where $\vec{x}_0 \in \ell$ is chosen so that the vector $\vec{x}_0 - \vec{q}$ is perpendicular to ℓ. This means that this vector is parallel to $\vec{n}$. It also means that $\vec{q} - \vec{x}_0$ is the orthogonal projection of $\vec{p} - \vec{q}$ (or, of $\vec{x}_0 - \vec{q}$ instead, for any point $\vec{x}_0 \in \ell$) onto $\vec{n}$. Therefore we obtain that

$$\begin{aligned} d(\vec{q}, \ell) &= \|\mathrm{Proj}_{\vec{n}}(\vec{p} - \vec{q})\| \\ &= \frac{(\vec{p} - \vec{q}) \cdot \vec{n}}{\vec{n} \cdot \vec{n}} \|\vec{n}\| \\ &= \frac{(\vec{p} - \vec{q}) \cdot \vec{n}}{\sqrt{\vec{n} \cdot \vec{n}}}. \end{aligned}$$

In case of the plane in $\mathbb{R}^3$ the arguments as well as the formalism and the formula are the same! In fact in all dimensions, the distance $d(\vec{Q}, \Pi)$ between the point $\vec{Q} \in \mathbb{R}^n$ and the hyperplane Π is given by

$$d(\vec{Q}, \Pi) = \frac{(\vec{P} - \vec{Q}) \cdot \vec{N}}{\sqrt{\vec{N} \cdot \vec{N}}},$$

whenever $\vec{P}$ and $\vec{N}$ satisfy $\vec{P} \in \Pi$ and $\vec{0} \neq \vec{N} \perp \Pi$,

8.4.3 Triple product and the volume of a parallelepiped

Let $\vec{A}, \vec{B}$ and $\vec{C}$ are not coplanar in $\mathbb{R}^3$, which means that no plane passing through the origin generated by two vectors among them can contain the remaining vector. Then such three vectors generate a unique parallelepiped, say $\mathcal{P}$.

We are interested in explicating why $|\vec{A} \cdot \vec{B} \times \vec{C}|$ coincides with the volume of $\mathcal{P}$.

We have observed earlier that $\|\vec{B} \times \vec{C}\|$ coincides with the area of the parallelogram generated by two vectors $\vec{B}$ and $\vec{C}$. Now the volume of the parallelepiped $\mathcal{P}$ is the base area $\|\vec{B} \times \vec{C}\|$ times the height. Of course in this case the height is the length of the orthogonal projection of $\vec{A}$ onto $\vec{B} \times \vec{C}$. Therefore, we arrive at

$$\begin{aligned} \text{Volume of } \mathcal{P} &= \|\vec{B} \times \vec{C}\| \|\mathrm{Proj}_{\vec{B} \times \vec{C}}(\vec{A})\| \\ &= \|\vec{B} \times \vec{C}\| \left\| \frac{\vec{A} \cdot \vec{B} \times \vec{C}}{(\vec{B} \times \vec{C}) \cdot (\vec{B} \times \vec{C})} (\vec{B} \times \vec{C}) \right\| \\ &= |\vec{A} \cdot \vec{B} \times \vec{C}|. \end{aligned}$$

Chapter 9

Differentiability, partial derivatives, gradient

9.1 Functions $y = f(x_1, \ldots, x_n)$

The definition of a function is as follows:

Definition 9.1.1 A relation between X For sets X and Y, the concepts of *relation* and function are defined as follows. A *relation* is defined to be a subset of the Cartesian product $X \times Y$. A function $f\colon X \to Y$ is defined to be f satisfying the two conditions:

(1) There exists $y \in Y$ such that $(x, y) \in f$.

(2) $y_1 = y_2$ whenever $(x, y_1) \in f$ and $(x, y_2) \in f$.

In word, a *function* is a relation that *assigns a unique element in* Y *to each element of* X.

A function is an assignment of a unique value to each point in a certain set (called the domain of definition). In this course we usually consider a subset A of the set $\mathbb{R}^n$ of ordered n-tuples, say $(x_1, \ldots, x_n)$, of real numbers as the domain of the function f and the values are going to be in the set $\mathbb{R}$ of real numbers.

Then the classic notation $z = f(x_1, \ldots, x_n)$ also makes sense. This notation indicates that, for each point represented by $(a_1, \ldots, a_n)$ (in the set A), the value $f(a_1, \cdots, a_n)$ is computed by replacing each x_i by a_i in the formula $f(x_1, \ldots, x_n)$. amd the result is the assigned value for the function.

Before proceeding any further, we recall the traditional notation:

$$\|(x_1, \ldots, x_n)\| = \sqrt{x_1^2 + \ldots + x_n^2}$$

and with $\vec{a} = (a_1, \ldots, a_n)$, $\vec{x} = (x_1, \ldots, x_n)$

$$B^n(\vec{a}, r) := \{(x_1, \ldots, x_n) \in \mathbb{R}^n\colon \|\vec{x} - \vec{a}\| < r\}.$$

This is called the *open-ball* of with radius r centered at $\vec{a}$.

Our goal is to introduce Calculus (i. e., the differentiation and integration), for such functions of several variables, through the limits in all directions. Thus the domain U, say, of our functions better possess the property that any point of U can be approached from all possible directions, in U. The following concept provides such an environment.

Definition 9.1.2 A subset U of $\mathbb{R}^n$ is said to be *open* if, for every point p of U, there exists $r > 0$ such that $B^n(p, r) \subset U$.

So, it should be obvious now that, if a domain of the function in consideration is an *open* subset of $\mathbb{R}^n$, the current requirement is satisfied. Moreover, the graph of f is above its domain $U \subset \mathbb{R}^n$ and is a subset of $\mathbb{R}^{n+1}$, since we need one more axis, the y-axis. We shall see this in the popular cases such as $z = f(x, y)$ and $w = f(x, y, z)$ in what follows.

9.2 Differentiability—the case of $z = f(x,y)$

9.2.1 Graph

Let us start with the function $z = f(x,y)$, which assignes a unique value $f(x,y)$ to each point (x,y) in the xy-plane. Thus for each location at the point (x,y), one follows the direction parallel to the z-axis, which is the direction perpendicular to the xy-plane and record the value $z = f(x,y)$. So z-value is regarded as the height (or the depth, if z-value is negative) directly above the point described by the coordinate (x,y).

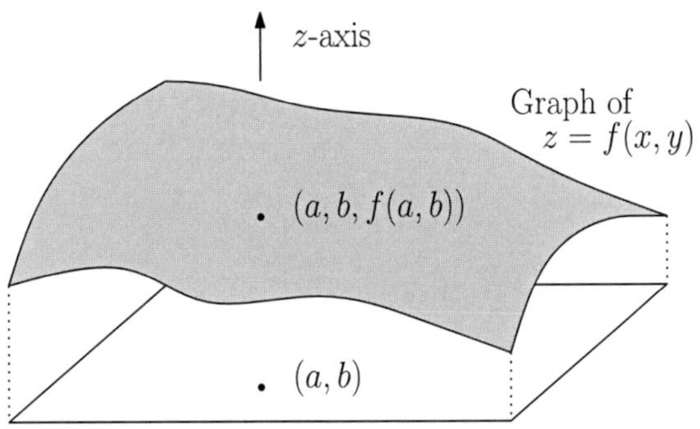

Figure 9.1: Graph of $z = f(x,y)$

9.2.2 Differentiability and Derivative

We are now ready to define the differentiability of a function $z = f(x,y)$ at a point (a,b).

Definition 9.2.1 (Differentiability of $f(x,y)$) Let $(a,b) \in \mathbb{R}^2$. If $z = f(x,y)$ is a function defined on the set $U := \{(x,y) \in \mathbb{R}^2 \colon \|(x,y) - (a,b)\| < r\}$ for some constant $r > 0$, then f is said to be *differentiable* at (a,b) if there exists $(m_1, m_2) \in \mathbb{R}^2$ such that

$$\lim_{\|(x-a,y-b)\|\to 0} \frac{f(x,y) - [f(a,b) + m_1(x-a) + m_2(y-b)]}{\|(x-a,y-b)\|} = 0. \tag{9.1}$$

Let us try to understand what this definition says. It may be helpful to go back to the case of $y = f(x)$. The differentiability at $x = a$ a point a on the real line $\mathbb{R}$ was defined by the existence of a real number m satisfying

$$\lim_{|x-a|\to 0} \frac{f(x) - [f(a) + m(x-a)]}{|x-a|} = 0. \tag{9.2}$$

Of course, most students (like me) learned the definition

$$\lim_{x\to a} \frac{f(x) - f(a)}{x - a} = m$$

but, it should not take long to realize that it is equivalent to (9.2).

On the other hand, (9.2) delivers a slightly different interpretation of derivative concept from the conventional one. The message in word is as follows: the differentiation m defines the line $\ell_a(x) = f(a) + m(x-a)$ that satisfies

$$\lim_{|x-a|\to 0} \frac{f(x) - \ell_a(x)}{|x-a|} = 0.$$

Namely, it says, "the line $y = \ell_a(x)$, when x is close by a, that is the closest to the function $y = f(x)$ can be determined." Of course, this, in mathematical terms, says that the function $f(x)$ is defined to be differentiable at a if the tangent line to $y = f(x)$ is uniquely determined at the point $(a, f(a))$.

In the same vein, Definition 9.2.1 can be understood now. Namely, the definition and the description (9.1) simply says that at the point $(a, b, f(a, b))$ there exists a plane passing through that point (given by the equation $z = L_{(a,b)}(x, y) = f(a, b) + m_1(x-a) + m_2(y-b)$) that approximates best the function $z = f(x, y)$ when (x, y) is close by (a, b). The following lemma may be helpful for the student:

Lemma 9.2.1 *Assume that a function $y = f(x, y)$ defined on the ball $U = B^2\big((a, b), r\big)$ of radius $r > 0$ centered at (a, b) is differentiable at (a, b). Show that the plane satisfying the condition (9.1) is unique.*

Proof. Assume that the planes represented by the equations

$$L^j(x, y) = f(a, b) + m_1^j(x-a) + m_2^j(y-b)$$

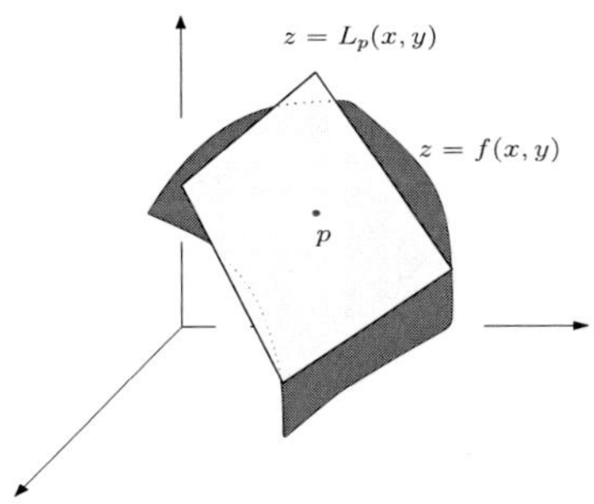

Figure 9.2: Tangent plane to $z = f(x, y)$

(where $j = 1, 2$) satisfy the condition (9.1). Then

$$0 = \lim_{\|(x-a,y-b)\|\to 0} \frac{f(x,y) - L^1(x,y)}{\|(x-a, y-b)\|} - \lim_{\|(x-a,y-b)\|\to 0} \frac{f(x,y) - L^2(x,y)}{\|(x-a, y-b)\|}.$$

Since both limits exist, this implies

$$\begin{aligned} 0 &= \lim_{\|(x-a,y-b)\|\to 0} \frac{L^2(x,y) - L^1(x,y)}{\|(x-a, y-b)\|} \\ &= \lim_{\sqrt{(x-a)^2+(y-b)^2}\to 0} \frac{(m_1^2 - m_1^1)(x-a) + (m_2^2 - m_1^1)(y-b)}{\sqrt{(x-a)^2 + (y-b)^2}}. \end{aligned}$$

This should hold in the particular case when $y = b$ and $x \to a$ while $x > a$. Therefore, we obtain

$$0 = \lim_{x\downarrow a} \frac{(m_1^2 - m_1^1)(x-a)}{x-a} = m_1^2 - m_1^1.$$

Similarly, consider the case when $x = a$ and $y \to b$ while $y > b$. Then we get $m_2^2 = m_2^1$. Altogether, one concludes that $L^1(x, y) = L^2(x, y)$, as desired. □

Therefore, it makes sense to call the plane defined by the expression $z = L_{(a,b)}(x, y) = m_1(x - a) + m_2(y - b) + f(a, b)$ satisfying the condition (9.1) the *tangent plane* to $y = f(x, y)$ at (a, b).

So, we close this section remarking that a function $z = f(x, y)$ is said to be *differentiable at* (a, b) if it admits the uniquely determined tangent plane at the point $(a, b, f(a, b))$.

Well then, what are those important values m_1, m_2 that determine the tangent plane? How can one compute them? We shall answer to the question in the next section.

Remark 9.2.1 The definition of limit is logically (formally) the same as the case of the one-variable function. By

$$\lim_{\vec{a}\to\vec{a}} f(\vec{x}) = L$$

we mean that, for any $\epsilon > 0$, there exists $\delta > 0$ such that

$$|f(\vec{x}) - f(\vec{a})| < \epsilon$$

for any $\vec{x}$ satisfying $0 < \|\vec{x} - \vec{a}\| < \delta$.

So the proof-technique and arguments are similar, as well. On the other hand, $\vec{x} = (x_1, \ldots, x_n) \in \mathbb{R}^n$ with $n > 1$ will require more thoughts in some aspects. Some examples are presented in Section 11.4.

9.3 Partial derivatives

Let us start now with the setting:

- $z = f(x, y)$ is defined on $B^2((a, b), r)$ and differentiable at (a, b).
- The tangent plane at (a, b) is represented by
$$z = m_1(x - a) + m_2(y - b) + f(a, b).$$

The first question to ask now is how to find the values for m_1, m_2. For this we return to (9.1). There we take the special case when $y = b$ and $x \to a$ while $x > a$, as before. What we see is

$$0 = \lim_{x \to a} \frac{f(x, b) - [f(a, b) + m_1(x - a)]}{x - a}$$

which implies

$$m_1 = \lim_{x \to a} \frac{f(x, b) - f(a, b)}{x - a}.$$

On the other hand, the term $\lim\limits_{x \to a} \dfrac{f(x, b) - f(a, b)}{x - a}$ is well-known as the *partial derivative of $f(x, y)$ with respect to the x-variable* at the point (a, b) and is denoted by

$$\left.\frac{\partial f}{\partial x}\right|_{(a,b)}.$$

In particular we have that $m_1 = \left.\dfrac{\partial f}{\partial x}\right|_{(a,b)}$. Similarly one immediately observes that the *partial derivative of $f(x, y)$ with respect to the y-variable at* (a, b)

$$\left.\frac{\partial f}{\partial y}\right|_{(a,b)} := \left.\frac{d}{dy}\right|_{y=b} f(a, y)$$

arises naturally to yield $m_2 = \left.\dfrac{\partial f}{\partial y}\right|_{(a,b)}$.

Altogether, we obtain that the tangent plane for $f(x, y)$ at $(a, b, f(a, b))$ is obtained by

$$z = L_{(a,b)}(x, y) = f(a, b) + (x - a)\left.\frac{\partial f}{\partial x}\right|_{(a,b)} + (y - b)\left.\frac{\partial f}{\partial y}\right|_{(a,b)} \tag{9.3}$$

This equation is the best approximation (as we already observed above) of $y = f(x, y)$ near $(x, y) = (a, b)$ by a linear function (i. e., a polynomial of degree at most 1), and hence the name *linearization of $z = f(x, y)$ at* $(x, y) = (a, b)$ is quite natural also.

9.4 Differentiabiltiy of $x_{n+1} = f(x_1, \ldots, x_n)$

Let n be a positive integer. Then one may consider the function

$$y = f(x_1, \ldots, x_n)$$

of n-variables and discuss the differentiability and related concept in a similar fashion.

Since a unique real value (for y) will be assigned to each value for the n-tuple $\vec{x} := (x_1, \ldots, x_n)$, the graph of the function f as well as the tangent *hyper*plane should be in the $n+1$ dimensional Euclidean space.

In case $n = 1$, it represents a straight line. If $n = 2$, then a plane in $\mathbb{R}^3$. When the dimension $n+1$ is more than 3, one can only imagine theoretically what might happen, thinking that it must be a certain "flat space" of dimension n—that is what this terminology *hyperplane* in $\mathbb{R}^{n+1}$ signifies.

Having that stated, we see that any hyperplane (not parallel to the x_{n+1}-axis) passing through the point $\big(a_1, \ldots, a_n, f(a_1, \ldots, a_n)\big)$ is defined by the equation

$$x_{n+1} = L_{\vec{a}}(\vec{x})$$

where

$$\begin{aligned} L_{\vec{a}}(\vec{x}) &= \vec{m} \cdot (\vec{x} - \vec{a}) + f(\vec{a}) \\ &= m_1(x_1 - a_1) + \cdots + m_n(x_n - a_n) + f(a_1, \ldots, a_n). \end{aligned} \tag{9.4}$$

Now the differentiability of $x_{n+1} = f(x_1, \ldots, x_n)$ is defined as follows:

Definition 9.4.1 (Differentiability) A function $f(\vec{x}) = f(x_1, \ldots, x_n)$ is said to be *differentiable at* $\vec{a} = (a_1, \ldots, a_n)$, if there exists $\vec{m} = (m_1, \ldots, m_n) \in \mathbb{R}^n$ satisfying the condition

$$\lim_{\|\vec{x}-\vec{a}\| \to 0} \frac{f(\vec{x}) - [f(\vec{a}) + \vec{m} \cdot (\vec{x} - \vec{a})]}{\|\vec{x} - \vec{a}\|} = 0. \tag{9.5}$$

As in the case of $z = f(x, y)$, the vector $\vec{m}$ here is uniquely determined as well.

Definition 9.4.2 The *i-th partial derivative of f at $\vec{a}$* $\in \mathbb{R}^n$ is defined to be

$$\left.\frac{\partial f}{\partial x_i}\right|_{\vec{a}} := \left.\frac{d}{dt}\right|_{t=a_i} f(\ldots, a_{i-1}, t, a_{i+1}, \ldots). \tag{9.6}$$

In particular, one sees that

$$m_i = \frac{\partial f}{\partial x_i}\Big|_{\vec{a}}$$

for every $i = 1, \ldots, n$.

Of course the *linearization* of f at $\vec{a}$ is given by

$$x_{n+1} = L_{\vec{a}}(\vec{x}) \tag{9.7}$$

given in (9.4) above, satisfying (9.5), namely,

$$\lim_{\|\vec{x}-\vec{a}\|\to 0} \frac{f(\vec{x}) - L_{\vec{a}}(\vec{x})}{\|\vec{x} - \vec{a}\|} = 0.$$

Therefore,

$$L_{\vec{a}}(\vec{x}) = f(\vec{a}) + \sum_{i=1}^{n}(x_i - a_i)\frac{\partial f}{\partial x_i}\Big|_{\vec{a}},$$

and the equation (9.7) represents the *tangent hyperplane* of f at $\vec{x} = \vec{a}$.

9.5 Partial derivatives again, and Class C^k

In the preceding section we introduced and studied the concept of partial derivative for the multivariable functions of the form $f(x_1, \ldots, x_n)$. Since they can be defined for the function f, even though f is not necessarily differentiable, it is a fair question to ask whether it is possible to tell the function f is differentiable at $\vec{a}$, for instance, by investigating its partial derivatives. In fact the following theorem is known.

Theorem 9.5.1 *If a real-valued function $f(x_1, \ldots, x_n)$ is defined in an open set in $\mathbb{R}^n$ containing a point $\vec{a}$, and if the partial derivative function $\frac{\partial f}{\partial x_i}$ for every $i \in \{1, \ldots, n\}$ is well-defined and continuous in an open ball of positive radius centered at $\vec{a}$, then f is differentiable at $\vec{a}$.*

Proof. We shall do the case of $n = 2$ only, leaving the general case as an exercise to the student.

Start with the following application of the mean value theorem:

$$\begin{aligned} f(x_1, x_2) - f(a_1, a_2) & \\ &= [f(x_1, x_2) - f(a_1, x_2)] + [f(a_1, x_2) - f(a_1, a_2)] \\ &= \left.\frac{\partial f}{\partial x_1}\right|_{(c_1, x_2)} (x_1 - a_1) + \left.\frac{\partial f}{\partial x_2}\right|_{(a_1, c_2)} (x_2 - a_2), \end{aligned}$$

where c_i is between x_i and a_i, for $i = 1, 2$. Since the first partial derivatives are (assumed to be) continuous, we obtain that

$$\begin{aligned} &\lim_{\|\vec{x}-\vec{a}\| \to 0} \frac{1}{\|\vec{x}-\vec{a}\|} \Big\{ f(x_1, x_2) \\ &\qquad - \Big[f(a_1, a_2) + \left.\frac{\partial f}{\partial x_1}\right|_{(a_1, a_2)} (x_1 - a_2) + \left.\frac{\partial f}{\partial x_2}\right|_{(a_1, a_2)} (x_2 - a_2) \Big] \Big\} \\ &= \lim_{\|\vec{x}-\vec{a}\| \to 0} \frac{1}{\|\vec{x}-\vec{a}\|} \Big\{ \Big(\left.\frac{\partial f}{\partial x_1}\right|_{(c_1, x_2)} - \left.\frac{\partial f}{\partial x_1}\right|_{(a_1, a_2)} \Big) (x_1 - a_1) \\ &\qquad + \Big(\left.\frac{\partial f}{\partial x_2}\right|_{(a_1, c_2)} - \left.\frac{\partial f}{\partial x_2}\right|_{(a_1, a_2)} \Big) (x_2 - a_2) \Big\}. \end{aligned}$$

We are only to show that this last is equal to zero.

To do this, for convenience, we shall use the short-handed notation:

$$f_i = \frac{\partial f}{\partial x_i}, \text{ for } i = 1, 2.$$

Since

$$\frac{|x_1 - a_1|}{\sqrt{(x_1 - a_1)^2 + (x_2 - a_2)^2}} \le \frac{|x_1 - a_1|}{\sqrt{(x_1 - a_1)^2}} = 1,$$

we have

$$\begin{aligned}
&\lim_{\|\vec{x}-\vec{a}\|\to 0} \tfrac{1}{\|\vec{x}-\vec{a}\|}\left\{\left(\tfrac{\partial f}{\partial x_1}\Big|_{(c_1,x_2)} - \tfrac{\partial f}{\partial x_1}\Big|_{(a_1,a_2)}\right)(x_1 - a_1)\right\} \\
&\quad \le \lim_{\|\vec{x}-\vec{a}\|\to 0} |f_1(c_1, x_2) - f_1(a_1, a_2)| \\
&\quad = 0,
\end{aligned}$$

owing to the fact that f_1 is continuous (at (a_1, a_2)) is used here again. The remaining part of the limit can also be shown to be zero, by a similar argument, which we leave to the student, as an exercise. □

Remark 9.5.1 The student should be able to see from

$$\frac{\partial f}{\partial x_i}(a_1, \ldots, a_n) = \frac{d}{dt} f(\ldots, a_{i-1}, t, a_{i+1}, \ldots)\Big|_{t=a_i}$$

that this partial derivative is simply the rate of infinitesimal change of the function f in the direction of the x_i-axis, from the "central" point at $\vec{a}$. We shall develope a bit further on this type of understanding in subsequent sections.

For future purposes, we give

Definition 9.5.1 The function $f(\vec{x})$ is said to be of *Class* C^k if all the mixed partial derivatives of order up to and including k, and if the k-th order partial derivatives are continuous. C^0 will mean the class of continuous functions. By C^∞ we mean *of class* C^k for every natural number k.

9.6 Chain rule

Theorem 9.6.1 (The chain rule) *If $f(x_1, \ldots, x_n)$ is differentiable at $\vec{a} = (a_1, \ldots, a_n)$, if $u_1(t), \ldots, u_n(t)$ are differentiable at $t = t_0$ with $u_i(t_0) = a_i$ for every $i = 1, 2, \ldots, n$, and if the partial derivatives of f are continuous, then*

$$\frac{d}{dt}\bigg|_{t=t_0} f(h_1(t), \ldots, h_n(t)) = \sum_{i=1}^{n} h_i'(t_0)\frac{\partial f}{\partial x_i}\bigg|_{\vec{a}}.$$

Proof. This actually follows by the mean value theorem, which yields that there exist $\hat{h}_i$, for each $i = 1, \ldots, n$, between $h_i(t_0)$ and $h_i(t)$ satisfying

$$\begin{aligned}
&f(h_1(t), \ldots, h_n(t)) - f(h_1(t_0), \ldots, h_n(t_0)) \\
&\quad = \big[f(h_1(t), \ldots, h_n(t)) - f(h_1(t_0), h_2(t), \ldots, h_n(t))\big] \\
&\qquad + \cdots \\
&\qquad\quad + \big[f(h_1(t_0), \ldots, h_{n-1}(t_0) + h_n(t)) - f(h_1(t_0), \ldots, h_n(t_0))\big] \\
&\quad = \tfrac{\partial f}{\partial x_1}\Big|_{(\hat{h}_1, h_2(t), \ldots, h_n(t))} (h_1(t) - h_1(t_0)) \\
&\qquad + \cdots + \tfrac{\partial f}{\partial x_1}\Big|_{(h_1(t_0), \ldots, h_{n-1}(t_0), \hat{h}_n)} (h_n(t) - h_n(t_0)).
\end{aligned}$$

Now the result follows by dividing by $t - t_0$ and taking $\lim_{t\to t_0}$. Notice that the continuity of the partial derivatives and the differentiability of $h_i(t)$ are crucial in this argument. □

Corollary 9.6.2 *If $f(x_1, \ldots, x_n)$ is of class C^1 at $\vec{a} = (a_1, \ldots, a_n)$ and if $h_1(t_1, \ldots, t_m), \ldots, u_n(t_1, \ldots, t_m)$ are differentiable at $\vec{\alpha} := (\alpha_1, \ldots, \alpha_m)$ with $h_i(\vec{\alpha}) = a_i$ for every $i = 1, 2, \ldots, n$, then*

$$\frac{\partial}{\partial t_k}\bigg|_{t=t_0} f(h_1(\vec{t}), \ldots, h_n(\vec{t})) = \sum_{i=1}^{n} \frac{\partial f}{\partial x_i}\bigg|_{\vec{x}=\vec{a}} \frac{\partial h_i}{\partial t_k}\bigg|_{\vec{t}=\vec{\alpha}}.$$

9.7 $\vec{v}$-derivatives and directional derivatives

9.7.1 $\vec{v}$-derivative

Let $f(\vec{x})$ be differentiable at $\vec{x} = \vec{a}$ then for any $\vec{v} \in \mathbb{R}^n$, the chain rule implies that

$$\frac{d}{dt}\bigg|_{t=0} f(\vec{a} + t\vec{v}) = \sum_{i=1}^{n} v_i \frac{\partial f}{\partial x_i}\bigg|_{\vec{a}}.$$

At this point, define by

$$\frac{\partial}{\partial \vec{v}} := \sum_{i=1}^{n} v_i \frac{\partial}{\partial x_i}$$

so that

$$\frac{\partial f}{\partial \vec{v}}\bigg|_{\vec{a}} = \sum_{i=1}^{n} v_i \frac{\partial f}{\partial x_i}\bigg|_{\vec{a}}.$$

This last represents the derivative of f at $\vec{a}$ along the vector $\vec{v}$. Thus this "operator" $\frac{\partial}{\partial \vec{v}}$ is called the *derivative along the vector* $\vec{v}$, or simply the *$\vec{v}$-derivative.*

9.7.2 Directional derivatives

For a nonzero vector $\vec{v}$, the unit vector $\hat{v} := \vec{v}/\|\vec{v}\|$ is usually called the *direction* (vector) of $\vec{v}$. Then, in many occasions, it is convenient to consider the $\hat{v}$-derivative

$$\mathrm{Dir}_{\vec{v}} f := \frac{\partial f}{\partial \hat{v}}.$$

This is called (in many Caculus books) the *directional derivative* of f in the direction of $\vec{v}$.

9.8 The gradient ∇f

Definition 9.8.1 The operator ∇ is defined to be

$$\nabla = \Big(\frac{\partial}{\partial x_1}, \dots, \frac{\partial}{\partial x_n}\Big)$$

which will be used for

$$\nabla f = \Big(\frac{\partial f}{\partial x_1}, \dots, \frac{\partial f}{\partial x_n}\Big),$$

called the *gradient* of f.

Such an abstract notation/definition naturally generate more notation as follows:

$$\begin{aligned}
\nabla f(\vec{a}) &:= \Big(\frac{\partial f}{\partial x_1}\Big|_{\vec{a}}, \dots, \frac{\partial f}{\partial x_n}\Big|_{\vec{a}}\Big) \\
\vec{v} \cdot \nabla &:= \sum_{i=1}^{n} v_i \frac{\partial}{\partial x_i} = \frac{\partial}{\partial \vec{v}} \\
\frac{\vec{v}}{\|\vec{v}\|} \cdot \nabla &:= \frac{\partial}{\partial \frac{\vec{v}}{\|\vec{v}\|}} = \mathrm{Dir}_{\vec{v}}
\end{aligned}$$

9.9 Geometry of Gradient and $\vec{v}$-derivative

9.9.1 Direction of the fastest ascent

Consider $z = f(x_1, x_2)$ defined on the open disc $D(\vec{a}, r) := \{\vec{x} = (x_1, x_2) \colon \|\vec{x} - a\| < r\}$ of radius r centered at $\vec{a} := (a_1, a_2)$. Then the line emanating from $\vec{a}$ in the direction of $\vec{v} := (v_1, v_2)$ with unit speed is represented by the parametrized equation

$$\vec{x}(t) = \vec{a} + t\frac{\vec{v}}{\|\vec{v}\|},$$

as discussed in the earlier sections. Now, the directional derivative $\mathrm{Dir}_{\vec{v}} f|_{\vec{a}} = D_{\frac{\vec{v}}{\|\vec{v}\|}} f_{\vec{a}}$ is the derivative $g'(0)$ of the function

$$g(t) := f(\vec{x}(t)).$$

Hence the directional derivative $\mathrm{Dir}_{\vec{v}} f|_{\vec{a}}$ is the rate of increase of the function f at $\vec{a}$ in the direction of $\vec{v}$.

Denote by θ the angle between $\nabla f|_{\vec{a}}$ and $\vec{v}$. Then, discussions of preceding section show

$$\mathrm{Dir}_{\vec{v}} f|_{\vec{a}} = \|\nabla f|_{\vec{a}}\| \cos\theta.$$

Therefore, $\mathrm{Dir}_{\vec{v}} f|_{\vec{a}}$ is the largest if $\theta = 0$. This implies

Proposition 9.9.1 *The gradient $\nabla f|_{\vec{a}}$ is in the direction the xy-plane along which the directional derivative is the largest. In other words, the gradient points the direction along which f increases the fastest.*

This is often said as: the gradient points the *direction of the fastest ascent.* I would like to point out again that the gradient $\nabla f|_{\vec{a}}$ is a vector in the xy-plane, whereas the graph of $z = f(x, y)$ (or, in our notation $z = f(x_1, x_2)$, and hence the domain is in the x_1x_2-plane) is in the xyz-space.

Needless to say, all these ideas and conventions are trivially generalized to all dimensions and $x_{n+1} = f(x_1, \ldots, x_n)$, in which case the gradient of f is a vector in the $\mathbb{R}^n$ with $x_1, \ldots, x_n$-axes.

9.9.2 The level curves/surfaces and the gradient

If one specifies the value of the function $f(x, y)$ as c, then the intersection of the graph, say G_f, of $z = f(x, y)$ and the horizontal plane $z = c$ is the curve on the graph G_f consisting of point with equal height c. Such a curve is called a *contour* (or a *level curve*). The contour is in fact

$$\gamma_c = \{(x, y, c) \colon c = f(x, y)\}.$$

The projection of this curve onto the xy-plane is the curve represented by the equation $c = f(x, y)$. Repeating the obvious, the function does not change its value when (x, y) moves along this projected contour in the xy-plane.

So it is natural to expect that the direction tangent to the projected contour should be different from the direction of the gradient, along which the value of the function increases the fastest. In fact, they are perpendicular to each other, and we shall check it. For this aim, we shall exploit the implicit function theorem again so that the contour is a smooth curve. Let $\vec{x}(t) = x(t)\vec{i} + y(t)\vec{j}$ be a smooth parametrization of the projected level curve γ_c, then

$$c = f((x(t), y(t)).$$

Differentiating, one obtains

$$\begin{aligned} 0 &= \frac{d}{dt} f((x(t), y(t)) \\ &= \frac{\partial f}{\partial x}\Big|_{\vec{x}(t)} \frac{dx}{dt}\Big|_t + \frac{\partial f}{\partial y}\Big|_{\vec{x}(t)} \frac{dy}{dt}\Big|_t \\ &= \nabla f|_{(x(t),y(t))} \cdot (x'(t)\vec{i} + y'(t)\vec{j}) \end{aligned}$$

by the Chain rule. This shows that the gradient of f is perpendicular to the tangent vector to the projected level curve.

Remark 9.9.1 Gradient $\nabla f(a, b)$ is not necessarily perpendicular to the tangent plane to the graph of $z = f(x, y)$ at $(a, b, f(a, b))$, although $\nabla f(a, b)$ is perpendicular to the level curve defined by $f(a, b) = f(x, y)$. On the other hand $\nabla(z - f(x, y))|_{(a,b,f(a,b))}$ is a vector in the xyz-space perpendicular to the the tangent plane to the graph of $z = f(x, y)$ at $(a, b, f(a, b))$, Explain why, in your own terms drawing pictures if necessary.

Chapter 10

Min-max principle for $z = f(x, y)$

10.1 Local extremum and saddle point

Motivated by the understanding of the local maxima and minima, one sees that there are basically 3 cases to consider for the function $f(x_1, \ldots, x_n)$ at each critical point.

Definition 10.1.1 (Critical point) If $z = f(\vec{x})$ is a differentiable function on an open set U in $\mathbb{R}^n$, then any point $\vec{a} \in U$ is called the *critical point* if $\nabla f|_{\vec{a}} = \vec{0}$, or equivalently the tangent plane is horizontal, represented by the equation $z = f(\vec{a})$.

Definition 10.1.2 (Local maximum, minimum, saddle) The value $f(\vec{a})$ for some $\vec{a} \in U$ is said to be a *local maximum*, if there exists a constant $r > 0$ such that

$$f(\vec{x}) \leq f(\vec{a}), \quad \forall x \in U \text{ with } \|\vec{x} - \vec{a}\| < r.$$

In such a case, we say that f *attains its local maximum at* $\vec{a}$.

Likewise, $f(\vec{b})$ for a point $\vec{b} \in U$ is called a *local minimum*, if there exists a constant $r > 0$ such that

$$f(\vec{x}) \geq f(\vec{a}), \quad \forall x \in U \text{ with } \|\vec{x} - \vec{a}\| < r.$$

Of course, in such a case, f is said to attain its local minimum at $\vec{b}$.

The critical point $\vec{p} \in U$ is called a *saddle point*, if $f(\vec{p})$ is neither a local maximum nor a local minimum.

10.2 The Taylor development

Consider $z = f(\vec{x})$ defined on an open set U in $\mathbb{R}^n$ containing the point $\vec{a} = (a_1, \ldots, a_n)$. Assume that f is of class $\mathcal{C}^{m+1}(U)$, i. e., it admits all mixed partial derivatives up to and including order $m+1$ with the $(m+1)$-th derivatives continuous at every $\vec{p} \in U$.

Then consider $g(t) = g_{(\vec{a},\vec{h})}(t) := f(\vec{a} + t\vec{h})$ where $\vec{h} = (h_1, \ldots, h_n) \in \mathbb{R}^n$. Since U contains the open unit ball $\mathbb{B}^n(\vec{a}, r)$ of radius r centered at $\vec{a}$, for some $r > 0$, $g(t)$ is well-defined on the interval $\{t \in \mathbb{R} \colon |t| < r'\}$ for some $r' > 0$.

If we use the notation

$$(\vec{h} \cdot \nabla)^k := \sum_{\substack{i_1, \ldots, i_n \geq 0 \\ i_1 + \ldots + i_n = k}} \frac{k!}{i_1! \cdots i_n!} h_1^{i_1} \cdots h_n^{i_n} \frac{\partial^{i_1 + \ldots + i_n}}{\partial x_1^{i_1} \cdots \partial x_n^{i_n}}$$

then the chain rule implies (one can easily show this by induction)

$$\left. \frac{d^k g}{dt^k} \right|_{t=0} = (\vec{h} \cdot \nabla)^k f|_{\vec{x} = \vec{a}}$$

for every $k = 1, \ldots, m$.

Now, we apply this to the case where $\vec{h} = \vec{x} - \vec{a}$. Then Taylor's theorem in one variable yields

$$\begin{aligned} f(\vec{a} + t(\vec{x} - \vec{a})) &= g_{(\vec{a}, \vec{x} - \vec{a})}(t) = g(t) \\ &= g(0) + \sum_{k=1}^{m} \frac{t^k}{k!} \left. \frac{d^k g}{dt^k} \right|_{t=0} + R_m \end{aligned}$$

where

$$R_m = \frac{t^{m+1}}{(m+1)!} \left. \frac{d^{m+1} g}{dt^{m+1}} \right|_{t=\tau}$$

for some τ between 0 and t. Rewriting this, we obtain the following theorem.

Theorem 10.2.1 (Taylor expansion/development) *If $f \in \mathcal{C}^{m+1}(U)$ for some open set U in $\mathbb{R}^n$ containing the point $\vec{a}$, then for any $\vec{x} \in \mathbb{B}(\vec{a}, r)$ satisfying $\varnothing \neq \mathbb{B}(\vec{a}, r) \subset U$, it holds that*

$$f(\vec{x}) - f(\vec{a}) = \sum_{k=1}^{m} \frac{1}{k!} \big((\vec{x} - \vec{a}) \cdot \nabla\big)^k f\big|_{\vec{a}} + E_m, \tag{10.1}$$

where

$$E_m = \frac{1}{(m+1)!}\left((\vec{x}-\vec{a})\cdot\nabla\right)^{m+1} f\big|_{\vec{\xi}}\,, \tag{10.2}$$

for some $\vec{\xi}$ on the line segment in $\mathbb{R}^n$ joining $\vec{a}$ and $\vec{x}$.

Remark 10.2.1 Altogether, the expansion formula can also be written as

$$\begin{aligned} f(x_1,\ldots,x_n) &= f(a_1,\ldots,a_n) \\ &\quad + \sum_{\substack{i_1,\ldots,i_n\geq 0 \text{ with} \\ k(:=i_1+\ldots+i_n)=1}}^{k=m} \frac{(x_1-a_1)^{i_1}\cdots(x_n-a_n)^{i_n}}{i_1!\cdots i_n!}\cdot\frac{\partial^k f}{\partial x_1^{i_1}\cdots\partial x_n^{i_n}}\bigg|_{(a_1,\ldots,a_n)} \\ &\quad + R_m \end{aligned}$$

where the remainder R_m takes the expression

$$R_m = \sum_{\substack{i_\ell\geq 0,\forall\ell=1,\ldots,n \text{ where} \\ i_1+\ldots+i_n=m+1}} \frac{(x_1-a_1)^{i_1}\cdots(x_n-a_n)^{i_n}}{i_1!\cdots i_n!}\cdot\frac{\partial^{m+1} f}{\partial x_1^{i_1}\cdots\partial x_n^{i_n}}\bigg|_{(\xi_1,\ldots,\xi_n)}$$

for some $\vec{\xi} = (\xi_1,\ldots,\xi_n)$ representing a point on the line segment joining $\vec{x} = (x_1,\ldots,x_n)$ and $\vec{a} = (a_1,\ldots,a_n)$.

10.3 Quadratic polynomials–aside

The second order term of the Taylor expansion of $z = f(x, y)$ at $\vec{a} := (a_1, a_2)$ is always a homogeneous polynomial of type

$$Q_2(x, y) = A(x - a_1)^2 + 2B(x - a_1)(y - a_2) + C(y - a_2)^2.$$

It is well-known that the positivity of such Q_2 is depending upon the sign of A and of the discriminant

$$\Delta := B^2 - AC.$$

Now we investigate (towards deciding whether the second order term has a definite sign) the cases as follows:

(1) If $A = 0$, then

$$\begin{aligned} Q_2(x, y) &= (y - a_2)\big(2B(x - a_1) + C(y - a_2)\big) \\ &= (y - a_2)(2Bx + Cy - 2Ba_1 - Ca_2) \end{aligned} \tag{10.3}$$

Now assume $A \neq 0$ from this point.

(2) If $\Delta < 0$, then

$$\begin{aligned} Q_2(x, y) &= \frac{1}{A}\Big(\big(A(x - a_1) + B(y - a_2)\big)^2 + (-\Delta)(y - a_2)^2\Big) \\ &= \frac{1}{A}\Big(\big(A(x - a_1) + B(y - a_2)\big)^2 + |\Delta|(y - a_2)^2\Big). \end{aligned} \tag{10.4}$$

(3) If $\Delta = 0$, then

$$\begin{aligned} Q_2(x, y) &= \frac{1}{A}\Big(A(x - a_1) + B(y - a_2)\Big)^2 \\ &= \frac{1}{A}\Big(A(x - a_1) + B(y - a_2)\Big)^2 \end{aligned} \tag{10.5}$$

(4) If $\Delta > 0$, then

$$\begin{aligned} &Q_2(x, y) \\ &= \frac{1}{A}\Big(\big(A(x - a_1) + B(y - a_2)\big)^2 - \Delta(y - a_2)^2\Big) \\ &= \frac{1}{A}\big(A(x - a_1) + (B + \sqrt{\Delta})(y - a_2)\big) \\ &\qquad \cdot \big(A(x - a_1) + (B - \sqrt{\Delta})(y - a_2)\big). \end{aligned} \tag{10.6}$$

10.4 The 2nd derivative test/criterion

Combining the discussions of two preceding sections, we arrive at the following criterion.

Theorem 10.4.1 (The 2nd Derivative Test) *Let* U *be an open subset of* $\mathbb{R}^2$ *and let* $\vec{a} = (a_1, a_2) \in U$ *be a critical point of a function* $z = f(x, y)$ *of class* $\mathcal{C}^2(U)$. *Then* $f(\vec{a})$ *is a local maximum, if the following two conditions hold:*

(1) $\Delta(\vec{a}, f) := \left(\left(\frac{\partial^2 f}{\partial x_1 \partial x_2} \right)^2 - \frac{\partial^2 f}{\partial x_1^2} \frac{\partial^2 f}{\partial x_2^2} \right) \Big|_{\vec{a}} < 0.$

(2) $\frac{\partial^2 f}{\partial x_1^2} \Big|_{\vec{a}} < 0.$

Likewise, $f(\vec{a})$ *is a local miminum, if the following two conditions hold:*

(1') $\Delta(\vec{a}, f) < 0.$

(2') $\frac{\partial^2 f}{\partial x_1^2} \Big|_{\vec{a}} > 0.$

Finally, if $\Delta(\vec{a}, f) > 0$, *then* $\vec{a}$ *is a saddle point of* f, *meaning that* $f(\vec{a})$ *is neither a local maximum nor a local minimum.*

Obviously, the theorem does not cover all possible cases; indeed this test remains inconclusive in the cases not mentioned in the theorem. On the other hand, similar results continue to hold in dimensions higher than 2, which we shall remark on in the next section.

10.5 Remarks on $z = f(x_1, \ldots, x_n)$

The second derivative criterion/test of the previous section was depending upon the cases when the term E_2 stays positive (negative, or changes sign, respectively) in sufficiently small an open ball centered at the critical point under consideration. Relying on the continuity of the 2nd order mixed partial derivatives of the function f, it is therefore crucial to be able to tell whether

$$Q_2(\vec{a}, f) = 2E_2 = \sum_{\alpha,\beta\in\{1,\ldots,n\}} \frac{\partial^2 f}{\partial x_\alpha \partial x_\beta}\bigg|_{\vec{a}} \cdot (x_\alpha - a_\alpha)(x_\beta - a_\beta)$$

shows the desired property with signs. That can be understood by the methods from Linear Algebra, especially the discussion on the determinants, eigenvalues as well as the positive definiteness of the Hessian matrix $\left(\frac{\partial^2 f}{\partial x_\alpha \partial x_\beta}\big|_{\vec{a}}\right)$, a symmetric $n \times n$ matrix. In fact the discussion in the preceding section is also part of it.

We do not intend to discuss this any more here, but it may be explained later should it become necessary. We shall at least mention a bit more how it can done in Appendix.

10.6 Lagrange multiplier method; Min/max problem with constraints

Find the maximum of the function $f(x_1, \ldots, x_n)$ when the variables $x_1, \ldots, x_n$ satisfies the equation $g(x_1, \ldots, x_n) = 0$.

Such a problem arise in real life. In case the value of the function indicates the price of a certain commercial item such as "ice-cream", the variables could take the value of the amount of production (supply, in terms of Economics), price of milk, temperature, air pressure (weather facts), religion spread (if there is one discouraging the purchase of ice-cream), latitude, ...). Then in many cases the variables are not completely independent, but they could in fact be related. The equation $g(x_1, \ldots, x_n) = 0$ above represents such relations or limitations. Thus it is rightly called the *constraint* equation. (It may come in the form of inequality, but we handle the exhibited case firest.)

Now we shall explicate the method by Lagrange in the case of two variables. The present goal is to answer to the following problem:

(†) *Find the maximum of the function* $z = f(x, y)$*, under the condition* $g(x, y) = 0$.

Lagrange's idea was that the curves represented by $c = f(x, y)$ evolves as the real number c increases. As discussed before, the value of the function increases (the fastest) along the direction of the gradient vector ∇f (in the xy-plane). So if c_3 were a local maximum (See the illustration below.) then the level curves will espcape leaving no intersection points with the curve represented by the equation $g(x, y) = 0$, as c becomes slightly larger than c_3.

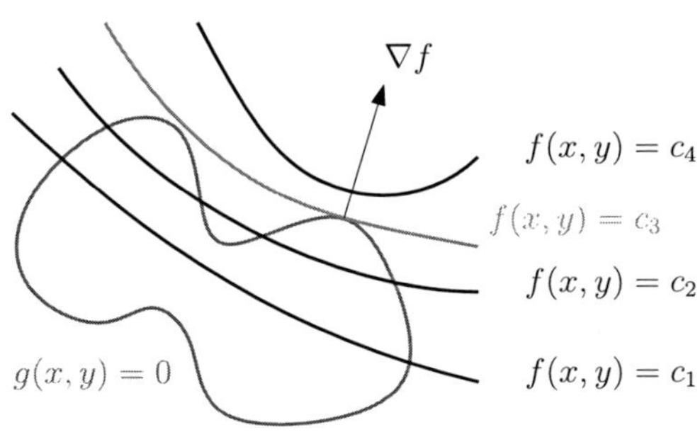

Figure 10.1: Lagrange's idea

On the other hand, if c is slightly smaller than the local maximum c_3, then there has to be a point (x, y) satisfying $g(x, y) = 0$ such that $f(x, y) = c$. So, it is highly plausible that the curve $c_3 = f(x, y)$ has to be tangent to the curve $0 = g(x, y)$. This means that one can find the local maxima and minima as follows.

In order to solve the mini/max problem (†) above, the Lagrange multiplier method proceeds along the following 3 steps:

Step 1. Solve for (x, y) (λ may be found as well, but unnecessary in the next steps) from the simultaneous equations

$$\begin{cases} \nabla f(x, y) = \lambda \nabla g(x, y) \\ g(x, y) = 0. \end{cases}$$

Step 2. At each solution (x, y) evaluate $f(x, y)$.

Step 3. The largest is the maximum, whereas the smallest is the minimum.

This method is called the *Lagrange multiplier method.*[1]

We exhibit a standard example, but then after that we shall discuss when this method is effective.

Example 10.6.1 *Find the maximum and the minimum of the function $f(x, y) = x^2 - 2y^2$ under the condition that $x^2 + y^2 = 1$.*

Remark 10.6.1 (Important!) In order for this Lagrange multiplier method to be effective, two conditions must be met.

(i) The constraint equation $g(x, y) = 0$ should define a closed differentiable curve. For the differentiability, it is sufficient if the gradient $\nabla g(x, y)$ exists and is continuous at every (x, y) satisfying $g(x, y) = 0$. But then the "closedness" of the curve is a bit tricky. At this level, it seems best to check it case by case.[2]

---

[1]Josephe-Louis Lagrange (1736-1813, France). The method is named as such because of the equation $\nabla f = \lambda \nabla$, namely the first gradient is a multiple of the second gradient.

[2]With a deeper understanding, it amounts to the condition that the set is a compact closed smooth curve. But that is also not so simple to check.

(ii) The function $f(x, y)$ has to be differentiable at every (x, y) satisfying $g(x, y) = 0$.

If these two conditions are satisfied, then the Lagrange multiplier method always works. If not. the method can still work in some cases, but in general there are cases where the method fails. See the following example.

Example 10.6.2 *Find the maximum and minimum for* $f(x, y) = x^2 - 2y^2$ *under the constraint* $x - y = 0$.

Remark 10.6.2 One may argue that the problem can be approached by solving for one of the variables in $g(x, y) = 0$ obtaining $y = h(x)$, for instance, and then the problem reduces to the min/max problem for $f(x, h(x))$. Such method is acceptable if it can be done. Even in higher dimensions! Nevertheless, Lagrange's method still has its own merit.

Remark 10.6.3 (Min/max problem on a bounded region) Sometimes the constraint comes in the form of inequalities, such as $x^2+y^2 \leq 1$. Then one first investigates the min/max for $f(x, y)$ in the interior of the surrounded region (the open unit disc represented by $x^2 + y^2 < 1$ in this case), and then the min/max problem can be treated by Lagrange's method. Combining the results, one solves the min/max problem on the bounded region given.

Now, we mention the Lagrange multiplier method in higher dimensions. The problem is simply finding the min/max for the function $f(x_1, \ldots, x_n)$ under the constraint $g(x_1, \ldots, x_n) = 0$. Then again the method is analogous: one solves the simultaneous equation

$$\begin{cases} \nabla f(x_1, \ldots, x_n) = \lambda \nabla g(x_1, \ldots, x_n) \\ g(x_1, \ldots, x_n) = 0 \end{cases}$$

and obtain the solutions, first. Then evaluate f at those points. The largest is the maximum and the smallest minimum. Of course the sufficient condition for this method to be effective, the gradient $\nabla g(x, y)$ has to exist and is nonzero, and the function f has better to be differentiable at every point satisfying $g(x_1, \ldots, x_n) = 0$.

Sometimes there are more than one constraints, such as $g_k(x_1, \ldots, x_n) = 0$ for $k = 1, 2, \ldots, m$. Then again consider

$$\begin{cases} \nabla f(x_1, \ldots, x_n) = \displaystyle\sum_{k=1}^{m} \lambda_k \nabla g_k(x_1, \ldots, x_n) \\ g_k(x_1, \ldots, x_n) = 0, \ \forall k = 1, 2, \ldots, m \end{cases}$$

and proceed analogously. We choose not to explain the sufficient conditions for this method to be efficient here, as it requires slightly (but not too much) more advanced an understanding of mathematics.

Chapter 11

Appendix: Some applications

11.1 Optimization/Linear programming

A convex polyhedron P in the xy-plane generated by the points $\vec{a}_0, \ldots, \vec{a}_m$ is given by

$$P = \{\vec{a}_0 + \sum_{k=1}^{m} r_k(\vec{a}_k - \vec{a}_0) \colon r_1, \ldots, r_m \geq 0, \sum_{k=1}^{m} r_k \leq 1\}.$$

Then the problem now is to find the maximum (or minimum) of the function

$$f(x, y) = ax + by$$

where a, b are the fixed constants and the point (x, y) traverses over P. This is a special case of the Lagrange multiplier problem, except that all the data as well as the function to investigate are linear.

In general, analogous problems exist in all dimensions. The polyhedron is replaced by the convex hull (the same definition, except that the ambient space is the higher dimensional $\mathbb{R}^n$ and the function to consider are of the form $f(x_1, \ldots, x_n) = b_1x_1 + \ldots + b_nx_n$.

In the spirit of Lagrange's multiplier method, notice first that $c = f(x, y)$ is a line. When c changes, lines move by a parallel transport. The nature of the polyhedron and the lines, the maximum or minimum occurs when the line touches only the boundary (but not the interior) in which case at least one of the generating points. Therefore, one can deduce the following solution method in steps:

- Step 1. Evaluate $f(\vec{a}_k)$ for all $k = 0, 1, \ldots, m$.
- Step 2. The largest is the maximum, etc. The point at which the maximum occurs is also found.

This solution method works in every dimension and any polytopes.

11.2 Least square method and y-Regression

When the set points $\mathcal{D} := \{(a_1, b_1), \ldots, (a_m, b_m)\}$ is given, it is often asked whether one can find a line that approximates these data the best. One way is finding the line $y = px + q$ where

$$E(p,q) = (b_1 - (a_1 p + q))^2 + \ldots + (b_m - (a_m p + q))^2$$

is the smallest. Here, finding the most optimal values for p, q is the goal/solution, making $E(p,q)$ possibly the smallest. If the solution is found, $y = px + q$ with the optimal $p.q$ is called the *y-regression* line.

This can be handled by what we learn from this course. The equations (simultaneous)

$$\begin{aligned} 0 &= \frac{\partial}{\partial p} E(p,q) = -2 \sum a_i (b_i - a_i p - q) \\ 0 &= \frac{\partial}{\partial q} E(p,q) = -2 \sum (b_i - a_i p - q) \end{aligned}$$

need to be solved in order for us to find the critical points of the function $E(p,q)$ in the variables p, q. Thus one obtains

$$\begin{aligned} p &= \frac{m \sum a_i b_i - \sum a_i \sum b_i}{m \sum a_i^2 - (\sum a_i)^2}, \\ q &= \frac{1}{m}\Big(\sum b_i - p \sum a_i\Big) \\ &= \frac{1}{m}\left(\sum b_i - \frac{m \sum a_i b_i - \sum a_i \sum b_i}{m \sum a_i^2 - (\sum a_i)^2} \sum a_i\right). \end{aligned} \tag{11.1}$$

So the function $E(p,q)$ has only one critical point, computed as such. Then we see also that

$$\lim_{p^2+q^2 \to \infty} E(p,q) = +\infty.$$

Consequently the minimum is attained for $E(p,q)$ at the critical point. Thus in this sense, one may accept that $y = px + q$ with p, q as in (11.1).

11.3 Graphs given by polar coordinates

The polar coordinate system for the xy-plane is given by $[r, \theta]$ where the relation between the Cartesian (x, y)-coordinates and the polar coordinates can be described by

$$r^2 = x^2 + y^2, \quad \theta = \tan^{-1} \frac{y}{x}$$

or,

$$x = r \cos \theta, \quad y = r \sin \theta.$$

As one immediately observes, polar coordinate system has ambiguity built in; the same point may receive infinitely many different representations. Despite, polar coordinate system is widely used as it has many other merits.

In this section, we shall describe some of the polar graphs given by the equations of the form $g(r, \theta) = 0$. To say it more precisely, we shall describe the set

$$\Gamma_g = \{(x, y) \colon g(r, \theta) = 0\}.$$

Example 11.3.1 *Describe, in the xy-plane, the set described by*

$$r = \cos \theta,$$

Solution 1. As we always did, we shall plot a few points and try to connect them. Start with $\theta = 0$. Then we get $[1, 0]$ right-a-way. Then $[\frac{\sqrt{3}}{2}, \frac{\pi}{6}]$, $[\frac{\sqrt{2}}{2}, \frac{\pi}{4}]$, $[\frac{1}{2}, \frac{\pi}{3}]$, $[0, \frac{\pi}{2}]$ are obtained. When the angle becomes "obtuse", i. e., $\frac{\pi}{2} < \theta < \pi$, then we get

$$[-\tfrac{1}{2}, \tfrac{\pi}{2} + \tfrac{\pi}{6}], [-\tfrac{\sqrt{2}}{2}, \tfrac{\pi}{2} + \tfrac{\pi}{4}], [-\tfrac{\sqrt{3}}{2}, \tfrac{\pi}{2} + \tfrac{\pi}{3}],$$

etc. Now we face a problem: what do we do with $r < 0$? A natural answer may be to accept it as

$$[-a, \theta] = [a, \pi + \theta].$$

This way allows us to consider all possible components[1] of the set Γ_g. Considering the special angles whose cosine values are well-known as such, we have the picture:

With only a few points it is difficult to see what kind of a curve it is. But in this case we can see a different solution.

[1]Some authors would like to stick to the restriction $r \geq 0$. But we shall take more conventional approach.

Solution 2. Use the conversion formula. Muliplying r to both sides, we get

$$r = 0 \text{ or } r = \cos\theta \ \Leftrightarrow r^2 = r\cos\theta.$$

This last is equivalent to

$$x^2 + y^2 = x$$

which in turn is the same as

$$(x - \tfrac{1}{2})^2 + y = (\tfrac{1}{2})^2.$$

Hence it is the circle of radius $\frac{1}{2}$ centered at $(0, \frac{1}{2})$. The graph passes through the origin. Thus, $r = 0$ is not an additional condition, as far as the graph set is concerned.

Example 11.3.2 *Describe* $\Gamma = \{(x, y)\colon r = 1 + \cos\theta\}$.

Solution. The second solution method above (by converting the equation to that of x, y variables) was nice when it worked. But the current one gives no particular help in that direction. (Please, give a try, to see why for yourself.) So we shall largely rely upon the primitive method 1. We plot the points such as

$$[2, 0], [1 + \tfrac{\sqrt{3}}{2}, \tfrac{\pi}{6}], [1 + \tfrac{\sqrt{2}}{2}, \tfrac{\pi}{4}], [1 + \tfrac{1}{2}, \tfrac{\pi}{3}], [1, \tfrac{\pi}{2}],$$
$$[1 - \tfrac{1}{2}, \tfrac{\pi}{2} + \tfrac{\pi}{6}], [1 - \tfrac{\sqrt{2}}{2}, \tfrac{\pi}{2} + \tfrac{\pi}{4}], [1 - \tfrac{\sqrt{3}}{2}, \tfrac{\pi}{2}\tfrac{\pi}{3}], [0, \tfrac{\pi}{2} + \tfrac{\pi}{2}].$$

Then the mirro image (with respect to the x-axis) is the rest of the graph (and repeats, as the angles continue to increase/decrease). This gives the famous “cardioid,” a heart-shaped curve.

We wish to remark that one may provide a bit finer information by analyzing the derivative of the equation at the origin and the point $[2, 0]$, at which $x = 2, y = 0$. Namely, from

$$x^2 + y^2 = (x^2 + y^2)^{\frac{1}{2}} + x$$

we shall compute $x' = \frac{dx}{dy}|_{(2,0)}$. Direct calculation yields

$$2xx' + 2y = \frac{xx' + y}{\sqrt{x^2 + y^2}} + x'.$$

Putting in $x = 2, y = 0$, one gets $4x' = 2x'$ which implies $\frac{dx}{dy} = 0$. Thus the curve is vertical at $[2, 0]$. So the “graph” has the following shape:

Then there are several interesting shapes. The following are left to the reader to draw the shapes of their *polar graphs*.

Exercise 11.3.1 *Describe the polar graphs, i. e., the set* Γ_g *for the equation* $g(r, \theta) = 0$*, where:*

(1) $g = 0 \Leftrightarrow r = 5 + 3\cos\theta$.

(2) $g = 0 \Leftrightarrow r = 3 + 5\cos\theta$.

(3) $g = 0 \Leftrightarrow r^2 = \sin\theta$.

(4) $g = 0 \Leftrightarrow r = \cos 3\theta$.

11.4 Remarks on limits for $f(x, y)$

The aim of this section is to discuss how to handle the limit $\lim_{(x,y)\to(p,q)} f(x, y)$. The definition appears relatively simple, given by so-called "ϵ-and-δ" statement. On the other hand, computation is still somewhat confusing especially with $\lim_{(x,y)\to(p,q)} f(x, y)$. The difficulty is that there are infinitely many different "paths" for (x, y) to approach (p, q) and yet, the limit value has to be the same regardless of the choices for the paths. Consider the following examples.

Example 11.4.1 *Investigate* $\lim_{(x,y)\to(0,0)} \frac{2xy}{x^2+y^2}$.

Take the line $ax - by = 0$ with $(a, b) \neq (0, 0)$. These represents all straight lines passing through the origin. Now,

$$\lim_{\substack{x\to 0\\ ax=by}} \frac{2xy}{x^2+y^2} = \lim_{t\to 0} \frac{2abt^2}{(a^2+b^2)t^2} = \frac{2ab}{a^2+b^2}.$$

This last changes values depending upon the choice for (a, b). Therefore the limit does not exist.

Example 11.4.2 *Investigate the limit* $\lim_{(x,y)\to(0,0)} \frac{xy^2}{x^2+y^4}$. *This limit is known not to exist; give an explication.*

A naive thinking would consider the radial paths toward the origin. Take $\vec{v} = (a, b) \neq (0, 0)$, and then take $\alpha_{\vec{v}}(t) = t\vec{v} = (at, bt)$. Then along the path $\alpha_{\vec{v}}(t)$ the limit would be computed as follows. If $a = 0$, then $b \neq 0$ and

$$\lim_{t\to 0} \frac{(at)(bt)^2}{(at)^2 + (bt)^4} = 0$$

and if $a \neq 0$, then

$$\lim_{t\to 0} \frac{(at)(bt)^2}{(at)^2 + (bt)^4} = \lim_{t\to 0} \frac{abt}{a^2 + b^4t^2} = 0,$$

regardless of the choice for the nonzero vector (a, b). At this point, can one conclude that the limit is 0?

But there are other paths. For instance, let us take the path $\beta(t);= (kt^2, t)$, with $k \neq 0$ and let $t \to 0$. Then

$$\lim_{(x,y)\to(0,0)} \frac{xy^2}{x^2+y^4} = \lim_{t\to 0} \frac{kt^4}{(k^2+1)t^4} = \frac{k}{k^2+1},$$

whose value changes according to the value of $k \neq 0$. Therefore, the limit does not exist.

Remark 11.4.1 Sometimes, polar coordinates can help. The condition that $(x,y) \to (0,0)$ is equivalent to the condition that $r^2 := x^2+y^2 \to 0$ which is in turn equivalent to $r \to 0$. Thus if one exploits the polar coordinates, the preceding limits can be viewed as follows:

$$\lim_{(x,y)\to(0,0)} \frac{2xy}{x^2+y^2} = \lim_{r\to 0} \frac{2r^2\cos\theta\sin\theta}{r^2} = 2\cos\theta\sin\theta.$$

So it is clear that the value of the limits varies according to the value of θ which determines the slopes $\frac{a}{b}$ for instance.

Then for the other limit, we see that

$$\lim_{(x,y)\to(0,0)} \frac{xy^2}{x^2+y^4} = \lim_{r\to 0} \frac{r^3\cos\theta\sin^2\theta}{r^2(\cos^2\theta + r^2\sin^4\theta)} = \lim_{r\to 0} \frac{r\cos\theta\sin^2\theta}{\cos^2\theta + r^2\sin^4\theta}.$$

Here, if one coniders the polar equation $r = \cos\theta$, it gives a curve passing through the origin. Along this path, the limit becomes

$$\lim_{r\to 0} \frac{r\cos\theta\sin^2\theta}{\cos^2\theta + r^2\sin^4\theta} = \lim_{r\to 0} \frac{\sqrt{1-r^2}}{2-r^2} = \frac{1}{2} \neq 0.$$

Thus, taking different (nonlinear path) here yields a different limit value, resulting in that the limt $\lim_{(x,y)\to(0,0)} \frac{2xy}{x^2+y^2}$ does not exist.

But not all examples are so complicated. Look at

Example 11.4.3 *Investigate* $\lim_{(x,y)\to(0,0)} \frac{x^2y}{3x^2+y^2}$.

Using the polar coordinates, one gets

$$\lim_{(x,y)\to(0,0)} \frac{x^2y}{3x^2+y^2} = \lim_{r\to 0} \frac{r\cos^2\theta\sin\theta}{3-2\sin^2\theta} = 0$$

since

$$\left|\frac{\cos^2\theta\sin\theta}{3-2\sin^2\theta}\right| \le \frac{1}{3-2|\sin^2\theta|} \le 1.$$

11.5 Curves and surfaces in $\mathbb{R}^n$

A curve in the space $\mathbb{R}^3$ (namely, the xyz-space) consists of points in $\mathbb{R}^3$ which forms the trace of a motion of a particle/point in the same space. This means that an equation assigning a point $\vec{r}(t) := x(t)\vec{i}+y(t)\vec{j}+z(t)\vec{j}$ to each value of t represents a curve, where $x(t), y(t), z(t)$ are functions of t. In Calculus, it is natural to consider the *differentiable* (smooth) curves which means that each component function $x(t), y(t), z(t)$ are differentiable.

We shall not distinguish between curves and motions. In this regard, the "parameter" t is regarded as time variable (even though we do not put any physical meaning to it). The vector value function $\vec{v}(t) = \vec{r}'(t)$ is called the *velocity* or the *tangent vector*. Then $\|\vec{v}(t)\|$ is sometimes called the *speed*.

Sometimes a surface, say S, is given by an equation $f(x, y, z) = 0$. In such a case, the condition that a curve $\vec{r}(t)x(t)\vec{i} + y(t)\vec{j} + z(t)\vec{j}$ is on the surface S receives the condition in the form of an equation, which is

$$0 = f(x(t), y(t), z(t)) = f(\vec{r}(t)).$$

Of course, there is a well-developed theory (called Differential Geometry) of such curves. But we shall not discuss any further here.

11.6 Graphing $z = f(x, y)$ and $f(x, y, z) = 0$ by level curves

Describing a terrain is an old art; people have been doing it using the level curves (i. e., contours). We shall use the same old method in describing the graph of $z = f(x, y)$. We shall go first by an example.

Example 11.6.1 *Graph $x^2 + y^2 - z^2 = 1$.*

The level curve at the altitude z is given by

$$x^2 + y^2 = (\sqrt{z^2 + 1})^2.$$

This is a circle in the xy-plane (in fact, it is on the plane parallel to the xy-plane with the altitude z) of radius $\sqrt{z^2 + 1}$ centered at the origin $(0, 0)$. The smallest possible radius is 1 at the level $z = 0$, and then, when $|z|$ is very large, the radius increases like $|z|$ (because 1 is relatively ignorable then). Thus there is a cone with apex at the origin with axis of symmetry on the z-axis and aperture $45°$, as the asymptotics, and the surface/graph of $x^2 + y^2 = (\sqrt{z^2 + 1})^2$ is the connected hyperboloid including the cone inside and approaching the cone asymptotically, whose horizontal sections are circles of radius $\sqrt{z^2 + 1}$ on each level of altitude z.

If the equation shows symmetry, then it could be convenient to use it.

Example 11.6.2 *Describe the graph $z = x^2 + y^2$.*

If we use the polar coordinates $[r, \theta]$ on the xy-plane, then the equation is just $z = r^2$. This is a parabola. Now, the equation has free translation property along the θ direction, which is the rotational symmetry around the z-axis. Thus the graph of $z = x^2 + y^2$ is the *paraboloid* which is the rotation of the parabola, say $z = x^2, y = 0$ about the z-axis.

Using one of these two methods, one can describe the graphs of many basic functions such as: $z = x^2 - y^2$, $z^2 = x^2 + y^2$ etc. We will stop here, but the student might like to explore many more examples.

Exercise 11.6.1 *Describe the graph of $z = 2xy$.*

Exercise 11.6.2 *Describe the graph of $z = x^2 - y^2$.*

Exercise 11.6.3 *Describe the graph of* $z = x^2 + 2xy + 3y^2$.

Exercise 11.6.4 *Describe the graph of* $z = (x - y)^2$.

One can classify the shapes of the quadratic functions in general such as:

$$z = Ax^2 + By^2 + Cxy + Dx + Ey + F$$

according to the conditions imposed on the coefficients $A, \ldots, F$. Then one may extend the theory to more general functions of the form $z = F(x, y)$. But we shall stop here.

Chapter 12

Integrals in $\mathbb{R}^n$

12.1 Integral in multidimensions

The concept and theory of multiple integrals, more precisely the multidimensional integrals, for the continuous functions here follows those of Riemann integral of functions of one variable.

Start with

$$A = (a_1, \dots, a_n) \in \mathbb{R}^n, B = (b_1, \dots, b_n) \in \mathbb{R}^n$$
$$E^n(A, B) = \{(x_1, \dots, x_n) \in \mathbb{R}^n \colon a_i \le x_i \le b_i, \forall i\}.$$

Let

$$D = \bigcup_{A,B}^{\text{finite}} E^n(A, B)$$

the union of some finitely many $E^n(A, B)$'s with the property that any two elementary sets have part of the boundaries in common (or no intersection at all). We shall call $E^n(A, B)$ an *elementary cells*, for a moment. The set D will be called an *elementary body*. We can see intuitively that any union of elementary cubes can be subdivided into a union of (still finitely many) elementary cubes any two of which have either no intersection or only part of their boundaries in common.

It is convenient to consider the concept of "mesh". If $\mathcal{C}$ is the set of finitely many elementary cells $E^n(A, B)$, then the *mesh* of $\mathcal{C}$ is defined by

$$m(\mathcal{C}) := \max\{|a_i - b_i| \colon E^n(A, B) \in \mathcal{C}\}.$$

Remark 12.1.1 In case $n = 2$, the construction above may be easier for the student to visualize. The elementary cell $E^2(A, B)$ is then a rectangle, and the body D a juxtaposition of such rectangles. If $n = 3$, then the elementary cells are rectangular boxes and D a body formed by the cells glued together.

Now, consider a continuous function $f(x_1, \ldots, x_n)$ with $f \geq 0$ defined on a set X containing a region R of $\mathbb{R}^n$ in such a way that there are elementary bodies $D(\mathcal{C})$ and $D(\mathcal{E})$ constructed by the collection of elementary cells $\mathcal{C}$ and $\mathcal{E}$, respectively, which satisfy the conditions

$$\mathcal{C} \subset \mathcal{E},$$
$$D(\mathcal{C}) \subset R \subset D(\mathcal{E}) \subset X.$$

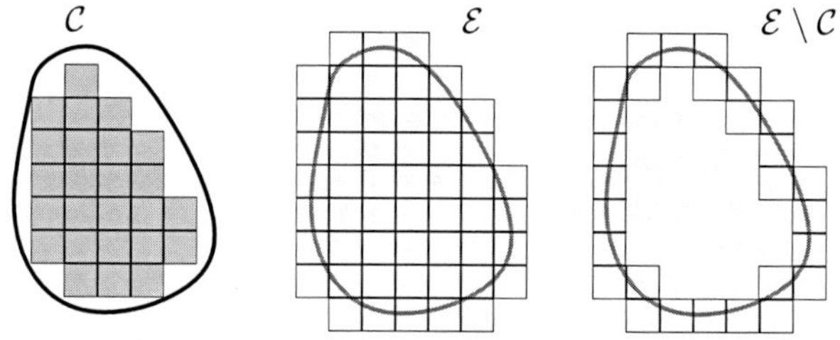

Then consider the points $x_{\min}, x_{\max} \in E^n(A, B)$ satisfying that $f(x_{\min}) = \min_{x \in E^n(A,B)} f(x)$ and $f(x_{\max}) = \max_{x \in E^n(A,B)} f(x)$ for each $E^n(A, B)$, respectively. Define by

$$L(f, \mathcal{C}) := \sum_{E^n \in \mathcal{C}} f(x_{\min})(b_1 - a_1) \cdots (b_n - a_n)$$
$$U(f, \mathcal{E}) := \sum_{E^n \in \mathcal{E}} f(x_{\max})(b_1 - a_1) \cdots (b_n - a_n)$$

Then

$$L(f, \mathcal{C}) \leq U(f, \mathcal{E}).$$

Moreover, we set

$$\begin{aligned} \|\mathcal{E} - \mathcal{C}\| &:= \sum_{E^n(A,B) \in \mathcal{E} \setminus \mathcal{C}} (b_1 - a_1) \cdots (b_n - a_n) \\ &= \text{Volume}\big(D(\mathcal{E}) \setminus D(\mathcal{C})\big). \end{aligned}$$

Proposition 12.1.1 *In the above setting, if R is a region such that* $\inf \|\mathcal{E} - \mathcal{C}\| = 0$, *then*

$$\sup_{\mathcal{C}} L(f, \mathcal{C}) = \inf_{\mathcal{E}} U(f, \mathcal{E}).$$

We shall not present the proof here, but it follows by the setting and the continuity. So we present

Definition 12.1.1 For the region R and a nonnegative function f satisfying the assumption as well as the conclusion in the proposition above, we call $\sup_{\mathcal{C}} L(f, \mathcal{C})$ the integral of f over the region R and denote by

$$\int_R f(x) dV := \inf_{\mathcal{E}} U(f, \mathcal{E}).$$

In case f changes sign, then we denote by:

$$f^+(x) = \begin{cases} f(x) & \text{if } f(x) \geq 0 \\ 0 & \text{if } f(x) < 0 \end{cases}$$

and $f^-(x) = (-f)^+$. Then,

$$f = f^+ - f^-,$$

and both f^+ and f^- are nonnegative continuous functions. Moreover, we shall define

$$\int_R f dV := \int_R f^+ dV - \int_R f^- dV.$$

Remark 12.1.2 Our definition of integral is for all possible finite dimensions. In case of dimension 2, this integral is the same as what is known as the *double integral.* In dimension 3, the triple integral. In general, these are called the *multiple integrals.* But I feel that the correct terminology should be *multidimensional integrals*, because of the exposition of the next section.

It is apparent that the regions R with condition

$$\inf \|\mathcal{E} - \mathcal{C}\| = 0$$

will play an important role in the study of mutidimensional integrals. Thus we shall call such regions *admissible regions.*

Of course, which regions are admissible is a significant question (not easy![1]), but we shall first deal with some examples, in order to gain some intuitive understanding, from the next few sections.

Remark 12.1.3 As is well known, the integrals can exist and be well defined for more general functions that are not necessarily continuous. But, for the purpose of learning the multivariable calculus of current level, this level of introduction seems adequate. For more information, the student may like to study higher level analysis.

[1]Interested reader may check on "Measure theory" books.

12.2 Evaluation by Iterated integrals

There always is a practical issue in every theory—and this time also; how does one actually compute the multiple integral, even though it exists? On a single (elementary cell $E^n(A, B)$, one would easily see that

$$\int_{E^n(A,B)} f(x_1, \ldots, x_n)dV = \int_{a_1}^{b_1} \Big(\cdots \Big(\int_{a_n}^{b_n} f(x_1, \ldots, x_n)dx_n \Big) \cdots \Big) dx_1.$$

The right-hand side is called the *iterated integral.* Saving the effort of drawing so many parentheses, we use instead the notation such as

$$\int_{a_1}^{b_1} \cdots \int_{a_n}^{b_n} f(x_1, \ldots, x_n)dx_n \cdots dx_1,$$

or equivalently

$$\int_{a_1}^{b_1} dx_1 \cdots \int_{a_n}^{b_n} f(x_1, \ldots, x_n)dx_n.$$

This viewpoint and method of evaluating multiple integrals extends to more general regions of integration, but we shall discuss this with more concrete examples, in the next section and what follows.

Remark 12.2.1 Turning multiple integrals into iterated integrals require a theorem which guarantees that the two objects are equal—whose more (if not the most) general form is known and Fubini-Tonelli theorem. But since we are dealing with the continuous functions for the integrand, full power of such theorem need not be drawn in our case.

12.3 Double integrals and polar coordinates

In this section, we shall deal with the integral with the region of integration in the xy-plane. So the "volume element" dV is written here as d^2X, dV^2 or dV_2, marking the dimension of the space.

12.3.1 By iterated integrals

We shall continue exposition with examples.

Example 12.3.1 *Evaluate* $\int_R e^x \sin y\, dV_2$ *where*

$$R = \{(x, y) \in \mathbb{R}^2 \colon |x| \le 1, 0 \le y \le \pi\}.$$

Solution. The iterated integrals representing the double integral are

$$\int_{-1}^{1} dx \int_0^{\pi} e^x \sin y\, dy \text{ and } \int_0^{\pi} dy \int_{-1}^{1} e^x \sin y\, dx.$$

A theorem by Fubini-Tonelli says that these two integrals coincide with each other, as they share the same value with the double integral $\int_R e^x \sin y\, dV_2$. But we shall evaluate the first iterated integral.

$$\begin{aligned}\int_R e^x \sin y\, dV_2 &= \int_{-1}^{1} dx \int_0^{\pi} e^x \sin y\, dy \text{ by Fubini-Tonelli theorem}\\ &= \int_{-1}^{1} e^x \big[-cosy\big]_0^{\pi}\, dx\\ &= \int_{-1}^{1} 2e^x\, dx\\ &= \big[2e^x\big]_{-1}^{1} = 2(e - e^{-1}).\end{aligned}$$

The second iterated integral gives the same answer as one sees:

$$\begin{aligned}\int_R e^x \sin y\, dV_2 &= \int_0^{\pi} dy \int_{-1}^{1} e^x \sin y\, dx \text{ by Fubini-Tonelli theorem}\\ &= \int_0^{\pi} \big[e^x\big]_{-1}^{1} \sin y\, dy\\ &= \int_0^{\pi} (e - e^{-1}) \sin y\, dy\\ &= 2(e - e^{-1}).\end{aligned}$$

Example 12.3.2 *The iterated integral* $A = \int_{-1}^{2} dx \int_{x^2}^{x+2} xy\ dy$ *is given. Answer to the following:*

(1) Describe the region of integration and then the double integral for A.

(2) Find the iterated integral in the other order.

Solution. For (1): The region, say R, of integration is described by y between x^2 and $x+2$, and by x between -1 and 2. Hence we conclude that

$$R = \{(x, y) \colon -1 \le x \le 2,\ x^2 \le y \le x + 2\}.$$

(Describing R pictorially may be helpful, but we leave it to the student.)

Now, for (2): In order to switch the order of iterated integral, we need to divide the region depending upon $0 \le y \le 1$ and $1 \le y \le 4$. Accordingly,

$$R = \{(x, y) \colon 0 \le y \le 1, -\sqrt{y} \le x\sqrt{y}\} \cup \{(x, y) \colon 1 \le y \le 4, 2 - y \le x \le \sqrt{y}\}.$$

Then we see that

$$A = \int_0^1 dy \int_{-\sqrt{y}}^{\sqrt{y}} xydx + \int_1^4 dy \int_{2-y}^{\sqrt{y}} xydx.$$

□

Example 12.3.3 *Evaluate the iterated integral* $A = \int_0^1 dy \int_y^1 e^{x^2}\ dx$.

Solution. Evaluating the first integral by dx is problematic. So we would like to investigate the situation by switching the order of integration. The region of integration is given by: y between 0 and 1, whereas x between y to 1. (Again, drawing the region of integration in the xy-plane is left to the student.) Thus the iterated integral in the other order becomes

$$A = \int_0^1 dx \int_0^x e^{x^2} dy.$$

This is better for computation. In fact we have,

$$\begin{aligned} A &= \int_0^1 dx \int_0^x e^{x^2} dy \\ &= \int_0^1 xe^{x^2} dx \\ &= \frac{1}{2}(e-1) \end{aligned}$$

□

12.3.2 By substitution; coordinate-change

Integration by substitution for the double integral is based upon mapping a region R into another region R^* via a transformation $\Phi(x, y) = (u, v)$ where:

(1) $\Phi(R) = R^*$

(2) x, y and u, v are related by

$$\begin{cases} u = u(x, y) \\ v = v(x, y). \end{cases}$$

We now introduce the concept of *Jacobian matrix* and *Jacobian determinant*, for the map $\Phi(x, y) = (u(x, y), v(x, y))$. The 2×2 matrix

$$J_\Phi := \begin{bmatrix} u_x & u_y \\ v_x & v_y \end{bmatrix}$$

is the Jacobian matrix of Φ, and its determinant

$$\mathcal{J}_\Phi = u_x v_y - u_y v_x$$

is the Jacobian determinant of Φ.

If $\Phi\colon R \to R^*$ is 1-1 and onto smooth map with smooth inverse $\Psi\colon R^* \to R$, then it is customary to denote by

$$\Phi(x, y) = (u(x, y), v(x, y)), \quad \Psi(u, v) = (x(u, v), y(u, v)).$$

We then present the following well-known theorem:

Theorem 12.3.1 *In the setting above, it holds that*

$$\int_R f(x, y)\ dxdy = \int_{R^*} f(x(u, v), y(u, v))\mathcal{J}_\Psi\ dudv,$$

where $dxdy$ denotes dV_2 in the xy-coordinates and $dudv$ does dV_2 in the uv-coordinates.

Deferring the explication for a moment, we might want to see an example first.

Example 12.3.4 *Evaluate* $\int_R e^{-x^2-y^2}\,dxdy$ *where* R *is the region given by the conditions* $x^2+y^2 \le 1$ *and* $y \ge 0$.

Solution. Let

$$r = \sqrt{x^2+y^2}, \theta = \tan^{-1}\frac{y}{x}.$$

This is the well-known polar coordinate system. Then let

$$\begin{aligned}\Phi(x,y) &= (r,\theta) = \left(\sqrt{x^2+y^2},\ \tan^{-1}\frac{y}{x}\right)\\ \Psi(r,\theta) &= (x,y) = (r\cos\theta, r\sin\theta).\end{aligned}$$

Then the Jacobian determinant $\mathcal{J}_\Psi$ is computed as follows:

$$\mathcal{J}_\Psi = x_r y_\theta - x_\theta y_r = \cos\theta\cdot r\cos\theta - r\cdot(-\sin\theta)\sin\theta = r.$$

Then what is the description of $R^* = \Phi(R)$ in terms of r,θ? Of course it is given by two inequalities $0 \le r \le 1, 0 \le \theta \le \pi$. Therefore, by Theorem 12.3.1 above, we have

$$\begin{aligned}\int_R e^{-x^2-y^2}\,dxdy &= \int_{R^*} e^{-r^2} r\cdot dr\,d\theta\\ &= \int_0^\pi d\theta\int_0^1 re^{-r^2}dr\\ &= \frac{\pi}{2}(1-e^{-1}).\end{aligned}$$

□

Changing the coordinates is a useful method, but it works better when a good coordinate system can be found. Otherwise it does not help much. We shall give one more example, nevertheless.

Example 12.3.5 *Let* $R = \{(x,y):\ -1 < 2x+y < 2, 1 < x-y < 2\}$. *Evaluate* $A = \int_R (2x+y)^3(x-y)^7\,dx\,dy$.

Solution. Let us take the coordinate change

$$\begin{aligned}\Phi(x,y) &= (u(x,y),v(x,y)) = (2x+y, x-y),\\ \Psi(u,v) &= (x(u,v),y(u,v)) = \left(\frac{1}{3}(u+v), \frac{1}{3}(u-2v)\right).\end{aligned}$$

Then $\mathcal{J}_\psi = -\frac{1}{3}$. So we get

$$A = \int_{-1}^{2} du \int_{1}^{2} u^3 v^7 \, dv$$

whose evaluation we leave to the student. □

An explication for Theorem 12.3.1

Before leaving this section, we discuss why Theorem 5.3.1 works. We use the notation for the transformations

$$(u,v) = \Phi(x,y), \ (x,y) = \Psi(r,\theta)$$

with

$$R^* = \Phi(R), \ \Phi \circ \Psi(u,v) = (u,v) \ \Psi \circ \Phi(x,y) = (x,y).$$

We only consider the case where $u(x,y), v(x,y), x(u,v), y(u,v)$ are differentiable functions.

Then, one can see that

$$\begin{aligned}
\Delta u \Delta v &\sim \left([(u_x\vec{i} + v_x\vec{j}) \times (u_y\vec{i} + v_y\vec{j})] \cdot \vec{k} \right) \Delta x \Delta y \\
&= (u_x v_y - u_y v_x) \Delta x \Delta y \\
&= \mathcal{J}_\Phi \Delta x \Delta y.
\end{aligned}$$

And, exchanging the roles, one also sees that

$$\Delta x \Delta y = \mathcal{J}_\Psi \Delta u \Delta v.$$

Of course, this is only a plausible and approximate reasoning. But we shall not go into any deeper in this direction. Interested reader should consult higher level books on mathematical analysis.

12.3.3 By Polar coordinates

As we saw from Example 12.3.4, the polar coordinates can be useful for many examples. One of the notable applications is as follows:

Example 12.3.6 *Evaluate the improper integral*

$$A = \int_{-\infty}^{\infty} e^{-x^2} dx.$$

The existence of this integral can be proved in several ways, comparing with x^{-2} for instance. So we shall skip its proof and simply use the fact. Then

$$\frac{A}{2} = \int_0^\infty e^{-x^2} dx = \lim_{b\to\infty} \int_0^b e^{-x^2} dx,$$

and

$$\begin{aligned}
\frac{A^2}{4} &= \Big(\lim_{b\to\infty} \int_0^b e^{-x^2} dx\Big)^2 \\
&= \lim_{b\to\infty} \Big(\int_0^b e^{-x^2} dx\Big)^2 \\
&= \lim_{b\to\infty} \Big(\int_0^b e^{-x^2} dx\Big)\Big(\int_0^b e^{-y^2} dy\Big) \\
&= \lim_{b\to\infty} \int_0^b dx \int_0^b e^{-x^2-y^2} dy \\
&= \lim_{b\to\infty} \int_{S_b} e^{-x^2-y^2} dV_2,
\end{aligned}$$

where $S_b = \{(x, y) \colon 0 \le x \le b, 0 \le y \le b\}$.

Let $R_s := \{(x, y) \colon x \ge 0, y \ge 0, \sqrt{x^2 + y^2} < s\}$ for $s > 0$. Then $R_b \subset S_b \subset R_{2b}$ for every $b > 0$. Since $e^{-x^2-y^2} > 0$, it follows that

$$\lim_{b\to\infty} \int_{R_b} e^{-x^2-y^2} dV_2 \le \lim_{b\to\infty} \int_{S_b} e^{-x^2-y^2} dV_2 \le \lim_{b\to\infty} \int_{R_{2b}} e^{-x^2-y^2} dV_2. \tag{12.1}$$

Now we compute $\int_{R_b} e^{-x^2-y^2} dV_2$ as in Example 12.3.4 here. With the polar coordinate system, we have

$$\begin{aligned}
x = r\cos\theta, \quad y &= r\sin\theta \\
dV &= r\ dr\ d\theta
\end{aligned}$$

and consequently

$$\begin{aligned}
\lim_{b\to\infty} \int_{R_b} e^{-x^2-y^2} dV &= \lim_{b\to\infty} \int_0^{\frac{\pi}{2}} d\theta \int_0^b re^{-r^2}\ dr \\
&= \lim_{b\to\infty} \frac{\pi}{4}(1 - e^{-b^2}) \\
&= \frac{\pi}{4}.
\end{aligned}$$

Similarly, we get

$$\lim_{b\to\infty}\int_{R_{2b}} e^{-x^2-y^2}dV_2 = \lim_{b\to\infty}\frac{\pi}{4}(1-e^{-(2b)^2}) = \frac{\pi}{4}.$$

These together with (12.1), we obtain that $\frac{A^2}{4}=\frac{\pi}{4}$ and hence

$$\int_{-\infty}^{\infty} e^{-x^2}dx = A = \sqrt{\pi}.$$

□

Example 12.3.7 *Evaluate the iterated integral*

$$A = \int_0^2 dx \int_{-\sqrt{4-x^2}}^{\sqrt{4-x^2}} \cos\frac{\pi}{8}(4-x^2-y^2)dy.$$

Solution. Note that

$$A = \int_R 2\cos\left(\tfrac{\pi}{8}(4-x^2-y^2)\right) dV_2,$$

where R is the right half of the circular disc of radius 2 centered at the origin in the xy-plane. If we use the polar coordinate system, then this double integral turns to

$$A = \int_{-\frac{\pi}{2}}^{\frac{\pi}{2}} d\theta \int_0^2 \cos\left(\tfrac{\pi}{8}(4-r^2)\right)\cdot 2r\ dr = 8.$$

Verification of the last equality is left to the student. □

Area integral

On an admissible region R in the plane $\mathbb{R}^2$, the integral $\int_R dV_2$ represents the quantity that can be interpreted as the volume of the tube over R with height 1, or equivalently the area of R. This yields:

$$\text{Area }(R) = \int_R dV_2 = \int_R dx\ dy = \int_R r\ dr\ d\theta.$$

Example 12.3.8 *Compute the area of the region R surrounded by the cardioid $r = 1+\cos\theta$.*

Solution. Let the area in question be A. Then

$$\begin{aligned} A &= \int_0^{2\pi} d\theta \int_0^{1+\cos\theta} r\ dr \\ &= \int_0^{2\pi} \frac{1}{2}(1+\cos\theta)^2 d\theta \\ &= \int_0^{2\pi} \frac{1}{2}\big(1 + 2\cos\theta + \frac{1}{2}(1+\cos 2\theta)\big) d\theta \\ &= \frac{3}{2}\pi. \end{aligned}$$

□

More generally, the area of the region described by

$$R = \{(r,\theta) \colon 0 \le r \le f(\theta), \theta_0 \le \theta \le \theta_1\}$$

is then given by

$$\begin{aligned} \text{Area }(R) &= \int_R dV_2 \\ &= \int_{\theta_0}^{\theta_1} d\theta \int_0^{f(\theta)} r\ dr \\ &= \frac{1}{2}\int_{\theta_0}^{\theta_1} (f(\theta))^2\ d\theta. \end{aligned}$$

Example 12.3.9 *Compute the area of the region R bounded by the graph of $r^2 = \sin 2\theta$ and with the half-lines $\theta = 0$ and $\theta = \frac{\pi}{2}$.*

Solution. The function to consider is $r = f(\theta) = \sqrt{\sin 2\theta}$. This is well-defined in the range $0 \le \theta \le \frac{\pi}{2}$. Then

$$\begin{aligned} \text{Area }(R) &= \int_0^{\frac{\pi}{2}} (f(\theta))^2 d\theta \\ &= \int_0^{\frac{\pi}{2}} \sin 2\theta\ d\theta \\ &= -\frac{1}{2}\cos 2\theta\Big|_0^{\frac{\pi}{2}} \\ &= 1. \end{aligned}$$

□

12.4 Triple integrals and polar type coordinates

Following the usual exposition of current multivariable calculus, we shall discuss Triple integrals, which is nothing but the 3 dimensional case of multiple (multidimensional!) integrals. Again we shall go into exposition focusing more into examples as in the 2 dimensional case.

12.4.1 Evaluations and their methods

As mentioned (generally) in Section 12.1, the existence has been treated for all (finite) dimensions. (We shall denote by dV_n the volume element in $\mathbb{R}^n$.) Thus in this section, we focus into the examples concerning the evaluation of the triple integrals $\int_D f\, dV_3$, of course by iterated integrals (relying upon the Fubini-Tonelli theorem that works in all finite dimensions).

Example 12.4.1 *Evaluate the integral* $A = \int_D xz^2 dV_3$, *where*

$$D = \{(x, y, z) \colon 0 \le x \le 2, \ -1 \le y \le 1, \ 0 \le z \le 5, \ x^2 \le z\}.$$

Solution. Notice that the range for z-variable is simply $x^2 \le z \le 5$. Therefore one of the natural orders of interated integrals seems

$$A = \int_{-1}^{1} dy \int_{0}^{2} dx \int_{x^2}^{5} xz^2 dz.$$

The detailed computation seems routine and henc is left to the student. □

We would like to remark that changing the order of the integral is also possible; there are 3! ways in all, of course. One, for instance, is

$$A = \int_{-1}^{1} dy \int_{0}^{4} dz \int_{0}^{\sqrt{z}} xz^2 dx + \int_{-1}^{1} dy \int_{4}^{5} dz \int_{0}^{2} xz^2 dx.$$

We shall leave it to the student if they want to try all possible orders of iterated integrals.

As in the case of double integrals, changing the order of iterated integrals sometimes provides an edge in evaluating the integral. (Then of course, it can create disadvantages!)

Example 12.4.2 *Evaluate*

$$A = \int_0^4 dz \int_0^1 dy \int_{2y}^2 \frac{2e^{x^2}}{\sqrt{z}}\, dx.$$

Solution. The first of iterated integral is difficult. So we try to change the order of integration via dz first, dx next and then dy the last. Then since the solid region is given by

$$2y \le x \le 2, \quad 0 \le y \le 1, \quad 0 \le z \le 4.$$

This condition describes the same region with

$$0 \le y \le \frac{x}{2}, \quad 0 \le x \le 2, \quad 0 \le z \le 4.$$

(Draw the picture in the xyz-space; this will help in what follows.) Hence,

$$\begin{aligned} A &= \int_0^4 dz \int_0^1 dy \int_{2y}^2 \frac{2e^{x^2}}{\sqrt{z}}\, dx \\ &= \int_0^4 dz \int_0^2 dx \int_0^{\frac{x}{2}} \frac{2e^{x^2}}{\sqrt{z}}\, dy \\ &= \int_0^4 dz \int_0^2 \frac{xe^{x^2}}{\sqrt{z}}\, dx \\ &= \int_0^4 z^{-\frac{1}{2}} \left[\tfrac{1}{2}e^{x^2}\right]_0^2 dz \\ &= \frac{e^4-1}{2} \int_0^4 z^{-\frac{1}{2}} dz \\ &= 2(e^4-1). \end{aligned}$$

□

12.4.2 Cylindrical coordinates

For the xyz-space (i.e., $\mathbb{R}^3$), there are two typical coordinate systems built upon the polar coordinate system for the xy-space (i.e., $\mathbb{R}^2$). We shall introduce the *cylindrical coordinate system* here, whereas the other one will be discussed in the next section.

The cylindrical coordinate system (r, θ, z) is given by[2]

$$\Phi := \begin{cases} r = \sqrt{x^2 + y^2} \\ \theta = \tan^{-1} \frac{y}{x} \\ z = z \end{cases} \qquad \Leftrightarrow \qquad \Psi(= \Phi^{-1}) := \begin{cases} x = r\cos\theta \\ y = r\sin\theta \\ z = z. \end{cases}$$

With the (r, θ, z) coordinates, the variable z stays unchanged, and hence we have:

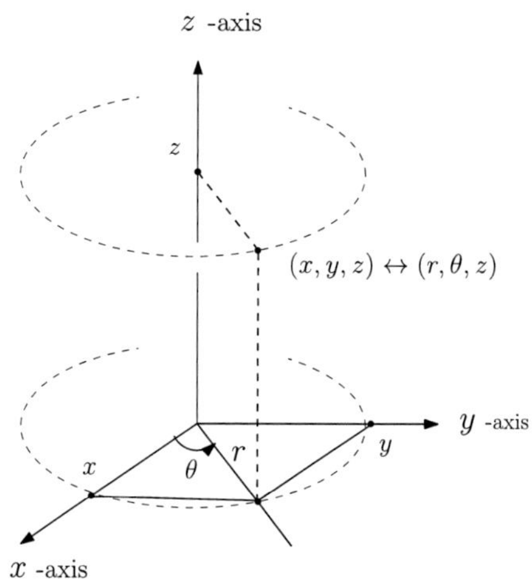

Figure 12.1: Cylindrical Coordinates

Theorem 12.4.1 *For a continuous function $f(x, y, z)$ defined on a set containing the admissible region R of integration, it holds that*

$$\int_R f \; dV_3 = \int_R f(x, y, z) \; dx \; dy \; dz = \int_{R^*} f(r\cos\theta, r\sin\theta, z) \; r \; dr \; d\theta \; dz,$$

where $R^* = \Phi(R)$.

[2]The value for r decides the cylinder of radius r centered built by taking the z-axis as the axis of rotation. Then θ-value decides the vertical line on the cylinder. Finally z-value decides where the point represented by (r, θ, z) on the line.

Of course there is a unified way of obtaining this. Namely we have the *Jacobian matrix* of Ψ

$$J_\Psi := \begin{pmatrix} x_r & x_\theta & x_z \\ y_r & y_\theta & y_z \\ z_r & z_\theta & z_z \end{pmatrix}$$

and the *Jacobian determinant* of Ψ

$$\mathcal{J}_\Psi := \det J_\Psi .$$

The determinant of 2×2 matix should be familiar to the student:

$$\det \begin{pmatrix} a_{11} & a_{12} \\ a_{21} & a_{22} \end{pmatrix} = \begin{vmatrix} a_{11} & a_{12} \\ a_{21} & a_{22} \end{vmatrix} := a_{11}a_{22} - a_{12}a_{21}.$$

Then the following inductive definition is convenient:

$$\begin{vmatrix} a_{11} & a_{12} & a_{13} \\ a_{21} & a_{22} & a_{23} \\ a_{31} & a_{32} & a_{33} \end{vmatrix} = (-1)^{1+1} a_{11} \begin{vmatrix} a_{22} & a_{23} \\ a_{32} & a_{33} \end{vmatrix} + (-1)^{1+2} a_{12} \begin{vmatrix} a_{21} & a_{23} \\ a_{31} & a_{33} \end{vmatrix} + (-1)^{1+3} a_{13} \begin{vmatrix} a_{21} & a_{22} \\ a_{31} & a_{32} \end{vmatrix}.$$

We shall present a more general treatment of the determinant of $n \times n$ matrices, later in Section 12.4.5, borrowing a context from Linear Algebra. But the exposition above should be sufficient to check that

$$\mathcal{J}_\Psi = \det J_\Psi = \begin{vmatrix} x_r & x_\theta & x_z \\ y_r & y_\theta & y_z \\ z_r & z_\theta & z_z \end{vmatrix} = \begin{vmatrix} \cos\theta & -r\sin\theta & 0 \\ \sin\theta & r\cos\theta & 0 \\ 0 & 0 & 1 \end{vmatrix} = r.$$

This is what the reader sees in Theorem 12.4.1. (Verify the last identity.)

Since the computation of integrals $\int_R f \, dV_3$ in the cylindrical coordinates have many features similar in principle to the case of the spherical coordinates which will be introduced in the next section, we shall present examples after the exposition of the next section (on the spherical coordinates).

12.4.3 Spherical coordinates

In dimension 3, there is one more natural generalization of polar coordinate system, which is

$$\Phi := \begin{cases} \rho = \sqrt{x^2+y^2+z^2} \\ \varphi = \cos^{-1}\frac{z}{\rho} \\ \theta = \tan^{-1} yx \end{cases} \quad \Leftrightarrow \quad \Psi(=\Phi^{-1}) := \begin{cases} x = \rho\sin\varphi\cos\theta \\ y = \rho\sin\varphi\sin\theta \\ z = \rho\cos\varphi, \end{cases}$$

where $0 \le \varphi \le \pi$. This is called the *spherical coordinate system.* Geometric explanation is as follows: First, the value for ρ describes the sphere of radius ρ centered at the origin. Secondly the value of $\varphi \in [0, \pi]$ describes the circle on the sphere with height $z = \rho\sin\varphi$. Then finally the value of θ decides which point on the circle it is.

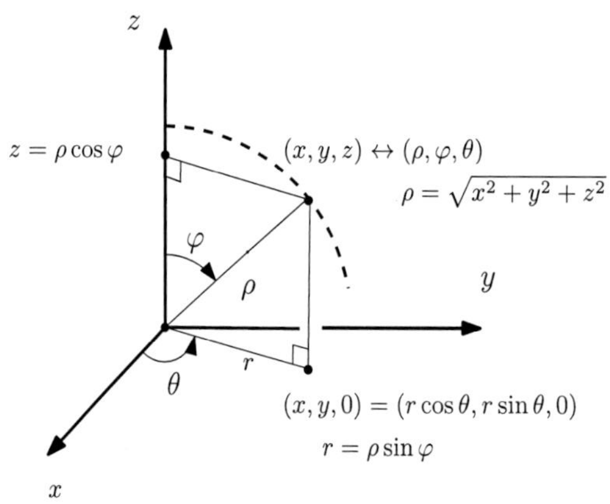

Figure 12.2: Spherical Coordinates

Now the Jacobian matrix and its determinant for the spherical coordinate system can be computed as follows:

$$\begin{aligned}
\mathcal{J}_\Psi &= \det J_\Psi \\
&= \begin{vmatrix} x_\rho & x_\varphi & x_\theta \\ y_\rho & y_\varphi & y_\theta \\ z_\rho & z_\varphi & z_\theta \end{vmatrix} \\
&= \begin{vmatrix} \sin\varphi\cos\theta & \rho\cos\varphi\cos\theta & -\rho\sin\varphi\sin\theta \\ \sin\varphi\sin\theta & \rho\cos\varphi\sin\theta & \rho\sin\varphi\cos\theta \\ \cos\varphi & -\rho\sin\varphi & 0 \end{vmatrix} \\
&= \rho^2\sin\varphi.
\end{aligned}$$

We leave the verification of the last two equalities to the student, again. On the other hand this yields:

Theorem 12.4.2 *With the spherical coordinate system (ρ, φ, θ) introduced above, and with R an admissible region for integration, it holds that*

$$\int_R f\ dx\ dy\ dz = \int_R f\ dV_3 = \int_{R^*} f\ \rho^2 \sin\varphi\ d\rho\ d\varphi\ d\theta,$$

where $R^* = \Psi(R)$.

12.4.4 Some examples of integrals

Example 12.4.3 *Evaluate $I_z = \int_D z\ dV_3$, where D is the solid region $D = \{(x, y, z)\colon 1 \le x^2 + y^2 + z^2 \le 4, z \ge 0\}$.*

Note: The center of mass of D of a uniform density 1 is at the point $\frac{1}{\text{Mass }(D)}(I_x, I_y, I_z)$, where $I_y = \int_D y\ dV_3$ and $I_z = \int_D z\ dV_3$.

Solution. The region D is the upper half of the difference between two spheres with common center at the origin and radii 2 and 1, respectively. Thus in spherical coordinates, the solid region can be represented by $1 \le \rho \le 2$, $0 \le \varphi \le \frac{\pi}{2}$ and $0 \le \theta \le 2\pi$. Thus

$$\begin{aligned}
I_z &= \int_0^{2\pi} d\theta \int_0^{\frac{\pi}{2}} d\varphi \int_1^2 \rho\cos\varphi \cdot \rho^2 \sin\varphi\ d\rho \\
&= \frac{1}{4}(2^4 - 1^4) \cdot 2\pi \cdot \int_0^{\frac{\pi}{2}} \sin\varphi\cos\varphi\ d\varphi \\
&= \frac{15\pi}{2}.
\end{aligned}$$

□

Example 12.4.4 *Compute the total mass of the solid D given by*

$$D = \{(x, y, z) \colon 0 \le z \le 4 - x^2 - y^2,\ 1 \le x^2 + y^2 \le 2\}$$

with density function $\delta(x, y, z) = 1 + z$.

Solution. The total mass equals the integral

$$A = \int_D (1 + z) dV_3.$$

We shall use the cylindrical coordinate system (r, θ, z). Then

$$\begin{aligned} A &= \int_0^{2\pi} d\theta \int_1^{\sqrt{2}} dr \int_0^{4-r^2} (1 + z) r \ dz \\ &= \int_0^{2\pi} d\theta \int_1^{\sqrt{2}} r\big((4 - r^2) + \tfrac{1}{2}(4 - r^2)^2\big) \ dr \\ &= \frac{17\pi}{3}. \end{aligned}$$

□

Example 12.4.5 *Evaluate the integral $A = \int_D dV_3$, where $D = \{(x, y, z) \colon x^2 + y^2 \le 1, y^2 + z^2 \le 1\}$.*

Solution.

$$\begin{aligned} A &= \int_{-1}^{1} dx \int_{-\sqrt{1-x^2}}^{\sqrt{1-x^2}} dy \int_{-\sqrt{1-x^2}}^{\sqrt{1-x^2}} dz \\ &= \int_{-1}^{1} (1 - x^2) dx \\ &= \frac{4}{3}. \end{aligned}$$

□

Example 12.4.6 *Evaluate the integral $A = \int_D dV_3$ where $D = \{(x, y, z) \colon z \ge \sqrt{x^2 + y^2}, x^2 + y^2 + z^2 \le 4\}$.*

Solution. Using the spherical coordinate system (ρ, φ, θ), we obtain

$$\begin{aligned} A &= \int_0^{2\pi} d\theta \int_0^{\frac{\pi}{4}} d\varphi \int_0^2 \rho^2 \sin\varphi \, d\rho \\ &= \frac{8}{3}\int_0^{2\pi} d\theta \int_0^{\frac{\pi}{4}} \sin\varphi d\varphi \\ &= \frac{8}{3}(2-\sqrt{2})\pi. \end{aligned}$$

□

Example 12.4.7 *Find the volume of the solid D bounded between two surfaces $z = 0$ and $z = x^2 + y^2$ and between two cylinders $(x-1)^2 + y^2 = 1$ and $(x-2)^2 + y^2 = 4$.*

Solution. We wish to use the cylindrical coordinate system (r, θ, z), even though we do not have to. Notice that

$$\begin{aligned} (x-1)^2 + y^2 = 1 &\Leftrightarrow r = 2\cos\theta \\ (x-2)^2 + y^2 = 4 &\Leftrightarrow r = 4\cos\theta. \end{aligned}$$

So,

$$\begin{aligned} \text{Volume } (D) &= \int_D dV_3 \\ &= \int_{-\frac{\pi}{2}}^{\frac{\pi}{2}} d\theta \int_{2\cos\theta}^{4\cos\theta} dr \int_0^{r^2} r \, dz \\ &= \int_{-\frac{\pi}{2}}^{\frac{\pi}{2}} d\theta \int_{2\cos\theta}^{4\cos\theta} \frac{1}{2} r^4 \, dr. \end{aligned}$$

This implies

$$\text{Volume } (D) = (4^4 - 2^4) \int_0^{\frac{\pi}{2}} \cos^4\theta \, d\theta = 45\pi.$$

Verifying the last equality is left to the student.[3] □

[3]This may be a good time point to recall "Wallis product" from calculus books or internet search.

12.4.5 On coordinate changes in n-variables, I

The n-dimensional Euclidean space can be coordinatized as

$$\mathbb{R}^n_x := \{(x_1, \ldots, x_n) \colon x_i \in \mathbb{R}, \forall i = 1, \ldots, n\}.$$

With another coordinate system, we write the first copy of

$$\mathbb{R}^n_u = \{\{(u_1, \ldots, u_n) \colon u_i \in \mathbb{R}, \forall i = 1, \ldots, n\}.$$

Then consider a smooth transformation $\Psi : \mathbb{R}^n_u \to \mathbb{R}^n_x$ and its coordinate representation

$$\Psi = \begin{cases} x_1 = x_1(u_1, \ldots, u_n) \\ \quad\vdots \\ x_n = x_n(u_1, \ldots, u_n). \end{cases}$$

Then its Jacobian determinant

$$\mathcal{J}_\Psi := \begin{vmatrix} \frac{\partial x_1}{\partial u_1} & \cdots & \frac{\partial x_1}{\partial u_n} \\ \vdots & \ddots & \vdots \\ \frac{\partial x_n}{\partial u_1} & \cdots & \frac{\partial x_1}{\partial u_n} \end{vmatrix}$$

takes an expression in $u_1, \ldots, u_n$.

Here, we need the formula for the determinant

$$\Delta(M) = \begin{vmatrix} a_{11} & \cdots & a_{1n} \\ \vdots & \ddots & \vdots \\ a_{n1} & \cdots & a_{nn} \end{vmatrix}$$

of the matrix

$$M = \begin{pmatrix} a_{11} & \cdots & a_{1n} \\ \vdots & \ddots & \vdots \\ a_{n1} & \cdots & a_{nn} \end{pmatrix}.$$

We shall discuss how to compute the determinant a moment later, and continue with the coordinate changes in n-dimensional integral. It is

$$\begin{aligned} &\int_D f(x_1, \ldots, x_n)\, dx_1 \cdots dx_n \\ &\qquad = \int_{D^*} f\big(x_1(\vec{u}), \ldots, x_n(\vec{u})\big) \mathcal{J}_\Psi \, du_1 \cdots du_n, \end{aligned} \tag{12.2}$$

where $D^* = \Psi^{-1}(D)$. This formula holds if Ψ is a smooth 1-1 map of $D^* \subset \mathbb{R}^n_u$ onto $D \subset \mathbb{R}^n_x$, with its inverse map also smooth. But then it is known that this change of variables formula is still valid for more general maps. But at this stage, we refrain from getting into any further in this direction.

Determinant $\Delta(M)$ of the matrix M

As written above, M is the $n \times n$ matrix whose entry at the i-th row and j-th column is the real number denoted by a_{ij}. We shall introduce two ways to define and compute its determinant $\Delta(M)$.

With $\mathbf{n} := \{1, 2, \ldots, n\}$, we consider the concept of *permutation* of n numbers, which is a one-to-one map of $\mathbf{n}$ onto itself. Note that, therefore, if $\sigma\colon \mathbf{n} \to \mathbf{n}$ is a permutation of n numbers, then

$$\sigma(1), \ldots, \sigma(n)$$

is simply a rearrangement of $1, \ldots, n$ with no repetition or deletion. Thus the terminology "permutation" appears reasonable. Denote by $\mathcal{S}_n$ the set of all permutations of n numbers.

Then a permutation $\tau \in \mathcal{S}_n$ is called a *transposition* if there are two distinct elements $i.j \in \mathbf{n}$ such that

$$\tau(i) = j, \tau(j) = i, \text{ and } \tau(k) = k, \forall k \in \mathbf{n} \setminus \{i, j\}.$$

Then the following is known:

Theorem 12.4.3 *Any permutation σ can be represented by the composition of transpositions, say $\sigma = \tau_1 \circ \tau_2 \circ \cdots \circ \tau_\ell$ for some integer $\ell \geq 0$. The decomposition is not in general unique, but whether ℓ is an even number or not is decided for each σ and remains unchanged.*

The permutation is called *even* if it can be represented by an even number of transpositions, and *odd* otherwise. Accordingly, the *sign* of a permutation is defined by

$$\operatorname{sgn}(\sigma) = \begin{cases} +1 & \text{if } \sigma \text{ is even} \\ -1 & \text{if } \sigma \text{ is odd.} \end{cases}$$

Then the *determinant* is defined as follows:

$$\Delta(M) = \sum_{\sigma \in \mathcal{S}} \operatorname{sgn}(\sigma) \cdot a_{1\sigma(1)} \cdots a_{n\sigma(n)}.$$

This definition allows us the inductive definition a special case of which was presented earlier in Section 12.4.2, after Theorem 12.4.1. For $i, j \in \mathbf{n}$, we define M_{ij} to be the $(n-1) \times (n-1)$ matrix obtained from M by removing the i-th row (i.e., the horizontal vector $(a_{i1}, \ldots a_{in})$) and the j-th column (consisting of the vertical vector with entries $a_{1j}, \ldots, a_{nj}$). This M_{ij} is usually called the *ij-th minor* of M. Then the inductive definition/formula (derived from the definition of determinant above) of the determinant is

$$\det M = \sum_{i=1}^{n} (-1)^{i+j} a_{ij} \det M_{ij} = \sum_{j=1}^{n} (-1)^{i+j} a_{ij} \det M_{ij}.$$

It may be helpful for the student if they try to evaluate the determinant of a 5×5 matrix with actual number entries.

12.4.6 On coordinate changes in n-variables, II

We feel that, even though we are not going to provide any rigorous explanation nor proofs, it will be beneficial to the student if we present some formalism for the change of coordinate systems.

The formalism we try to introduce of course has a solid conceptual meaning as well as precise theoretical foundation and developments. On the other hand, we shall simply introduce the computational technique cited from the big theory.

Differentials and 1-forms

The starting point is what is called the *differential* of a scalar valued function. That is, for a real-valued differentiable function $f(x_1, \ldots, x_n)$, the differential

$$df = \frac{\partial f}{\partial x_1} dx_1 + \ldots + \frac{\partial f}{\partial x_n} dx_n$$

which is called the *total differential* of the function f, in many other Multivariable Calculus books. In more general a viewpoint, we shall consider the objects

$$\alpha := g_1(x_1, \ldots, x_n) dx_1 + \ldots + g_n(x_1, \ldots, x_n) dx_n,$$

where $g_1, \ldots, g_n$ are smooth functions of n variables $x_1, \ldots, x_n$. Then students may like to ask what $dx_1, \ldots, dx_n$ are. Of course one may say that x_k

is the function defined by

$$x_k(a_1, \ldots, a_n) = a_k, \ \forall (a_1, \ldots, a_n) \in \mathbb{R}^n,$$

and dx_k is its (total) differential. Not quite satisfactory it is, we shall continue with these. The objects of type α, a linear combination of $dx_1, \ldots, dx_n$ with smooth function coefficients are going to be called *differential 1-form.*

The wedge product

We shall now introduce the binary operation $\wedge$ on 1-forms α and β which produces $\alpha \wedge \beta$, called a *2-form.* And we require that they satisfy the properties:

(1) $(\alpha \wedge \beta) \wedge \gamma = \alpha \wedge (\beta \wedge \gamma)$

(2) $\beta \wedge \alpha = -\alpha \wedge \beta$

(3) $(\alpha + \beta) \wedge \gamma = \alpha \wedge \gamma + \beta \wedge \gamma$

(4) $\big(f(\vec{x})\alpha\big) \wedge \big(g(\vec{x})\beta\big) = \big(f(\vec{x})g(\vec{x})\big)(\alpha \wedge \beta)$

for any smooth real-valued functions f, g and any 1-forms α, β, γ.

Then, by a *k-form,* we mean the k time wedge products of 1-forms, which are of the form

$$\sum_{i_1, \ldots, i_k \in \{1, \ldots, n\}} f_{i_1 \cdots i_k}(x_1, \ldots, x_n) dx_{i_1} \wedge \cdots \wedge dx_{i_k},$$

where $f_{i_1 \cdots i_k}(x_1, \ldots, x_n)$'s are smooth real-valued functions. Just formally, one sees therefore that if a k-form σ and an ℓ-form τ are "wedged", then the result $\sigma \wedge \tau$ turns out to be a $(k + \ell)$-form. The form $0 \ dx_1 \wedge dx_k$ is called the *null form* or simply *zero,* as adding this form does not change the result.

Notice that for any integer m with $m > n$, our construction shows that any m-form on $\mathbb{R}^n$ is zero!

Pullback operation by a smooth map of $\mathbb{R}^n$

Consider now a function (or, equivalently, mapping, map, or transformation) $\Psi \colon \mathbb{R}^n_{\vec{u}} \to \mathbb{R}^n_{\vec{x}}$. Then the map Ψ takes the values of the variables

$\vec{u} = (u_1, \ldots, u_n)$ to produce the values in $\mathbb{R}^n_{\vec{x}}$. Therefore, we see that Ψ receives the coordinate expression

$$\begin{aligned}(x_1, \ldots, x_n) &= \Psi(u_1, \ldots, u_n) \\ &= \big(\psi_1(\vec{u}), \ldots, \psi_n(\vec{u})\big).\end{aligned}$$

Then the mapping Ψ is called differentiable, if each component ψ_k is differentiable, for every k.

If we denote by $\Omega^\ell_{\vec{u}}$ the space of ℓ-forms on $\mathbb{R}^n_{\vec{u}}$, then it is known that there is a natural map $\Psi^*\colon \Omega^\ell_{\vec{x}} \to \Omega^\ell_{\vec{u}}$ defined by

$$\begin{aligned}\Psi^*(f_{i_1\cdots i_\ell}(\vec{x})dx_{i_1} \wedge \cdots \wedge dx_{i_\ell} + g_{j_1\cdots j_\ell}(\vec{x})dx_{j_1} \wedge \cdots \wedge dx_{j_\ell}) \\ = f_{i_1\cdots i_\ell}\big(\Psi(\vec{u})\big)d\psi_{i_1} \wedge \cdots \wedge d\psi_{i_\ell} \\ + g_{j_1\cdots j_\ell}\big(\Psi(\vec{u})\big)d\psi_{i_1} \wedge \cdots \wedge d\psi_{i_\ell}.\end{aligned}$$

Here of course,

$$d\psi_k = \sum_{i=1}^n \frac{\partial \psi_k}{\partial u_j} du_j$$

for every integer k, as we did for the total differential, earlier. This operation

$$\alpha \mapsto \Psi^*(\alpha)\colon \Omega^k_{\vec{x}} \to \Omega^k_{\vec{u}}$$

is called the *pullback* by Ψ.

All these must appeal to the student as overly abstract and makes them to ask whether it has any meaning! We need much patience; but we shall at least do some explicit computation of pullback operation at this stage.

Example 12.4.8 *Let $\Psi(u, v, w) = (uv, v + w, wu^2)$, and let $\alpha = y^2 dx \wedge dz - z dz \wedge dy$. Find an explicit expression of the pullback $\Psi^*(\alpha)$.*

Solution. By the above construction, we see that the expression $\vec{x} = \Psi(\vec{u})$ means

$$x = uv, \; y = v + w, \; z = wu^2.$$

Hence,

$$\begin{aligned}\Psi^*(y^2 dx \wedge dz) &= (v + w)^2 d(uv) \wedge d(wu^2) \\ &= (v + w)^2 (udv + vdu) \wedge (u^2 dw + 2uwdu) \\ &= u^2 v(v + w)^2 du \wedge dw + u^3 (v + w)^2 dv \wedge dw \\ &\quad - 2u^2 v(v + w)^2 du \wedge dv.\end{aligned}$$

Similarly,

$$\begin{aligned}\Psi^*(-zdz \wedge dy) &= -wu^2 d(wu^2) \wedge d(v+w) \\ &= (-2w^2u^3du - wu^4dw) \wedge (dv + dw) \\ &= -2w^2u^3du \wedge dv + wu^4dv \wedge dw - 2w^2u^3du \wedge dw.\end{aligned}$$

Altogether, we obtain

$$\begin{aligned}\Psi^*(\alpha) = &- \left(2u^2v(v+w)^2 + 2w^2u^3\right) du \wedge dv \\ &+ \left(u^3(v+w)^2 + wu^4\right) dv \wedge dw \\ &\quad + \left(u^2v(v+w)^2 - 2w^2u^3\right) du \wedge dw.\end{aligned}$$

□

Multiple integrals and pullback

Of course, the reason for all these formalism was for treating the coordinate change in multiple integrals. For this, we first observe that, for any n-form $\alpha \in \Omega^n_{\vec{x}}$ on $\mathbb{R}^n_{\vec{x}}$, there exists a scalar valued function

$$\alpha = f(\vec{x})\ dx_1 \wedge \cdots \wedge dx_n$$

for some smooth function $f(\vec{x})$. This follows by the construction of differential forms. We leave the checking to the student. Now we present

Definition 12.4.1 If R is an admissible region in $\mathbb{R}^n_{\vec{x}}$ and $f(\vec{x})$ is a continuous function then

$$\int_R f(\vec{x})dx_1 \wedge \cdots \wedge dx_n := \int_R f(\vec{x})dV_n.$$

Then the integration by substitution is treated by the following:

Theorem 12.4.4 *If R is an admissible region in $\mathbb{R}^n_{\vec{x}}$ and $f(\vec{x})$ is a continuous function, and if $\Psi\colon \mathbb{R}^n_{\vec{u}} \to \mathbb{R}^n_{\vec{x}}$ is a smooth mapping, then*

$$\int_R f(\vec{x})dx_1 \wedge \cdots \wedge dx_n = \int_{\Psi^{-1}(R)} \Psi^*\big(f(\vec{x})dx_1 \wedge \cdots \wedge dx_n\big).$$

Notice that the last is an integral in $\vec{u}$-coordinates.

Remark 12.4.1 Here, we remind the reader that

$$\Psi^{-1}(R) = \{p\colon \Psi(p) \in R\},$$

and so this definition need not require that the map Ψ should be 1-1. Therefore, this theorem is more general than the previous theorems on coordinate changes.

It should be a very good exercise for the student to check whether the coordinate change theorems with polar, cylindrical and spherical coordinates.

Chapter 13

Line and surface integrals

The present goal is to discuss the concept of integrals on the space other than part of Euclidean spaces. For this purpose the pullback discussed at the final part of the preceding chapter is particularly useful.

13.1 Line integral

13.1.1 Terminology and notation

We shall now discuss the line integral, an integral along a curve. By a *curve* in $\mathbb{R}^n$, we mean a function $\vec{r}(t)$ of an interval, say $[a, b]$ (with $a < b$) with values in $\mathbb{R}^n$. For our purpose, we shall primarily consider the curves that are differentiable, or at least piecewise C^1. We shall call these type of curves *admissible.*

As mentioned earlier in this lecture notes, such curves can be interpreted as a *motion* in $\mathbb{R}^n$, and has a *sense*, the "direction" of increasing t.

Now let D is an open region in $\mathbb{R}^n$ and let f a continuous real-valued function on D. Then let $\vec{r}$: $[a, b] \to D$ be a smooth (of class C^1) curve. More precisely, we say that a curve C is represented by the parametric equation $\vec{r}(t)$. In this setting, we denote by

$$ds := \|\vec{r}'(t)\|dt \tag{13.1}$$

and call it the *line element* of the curve $\vec{r}$. Then the line integral of f along the curve C parametrized by $\vec{r}(t)$ is denoted by

$$\int_C f\ ds$$

and is defined by

$$\int_C f\ ds := \int_a^b f(\vec{r}(t))\|\vec{r}'(t)\|dt. \tag{13.2}$$

Example 13.1.1 *Let $f(x,y) = x^2 - y^2$. A curve C in the xy-plane is parametrized by $\vec{r}(t) = (\cos t)\vec{i} + (\sin t)\vec{j}$, where $0 \le t \le \pi$. Evaluate the line integral $\int_C f\ ds$.*

Solution. Note that the curve is the circular path from $(1,0)$ to $(-1,0)$ along the upper half of the unit circle, with its sense counterclockwise. (But, there is no need to know such a geometric description for the evaluation of this line integral.)

By (13.1), $ds = \|(-\sin t)\vec{i} + (\cos t)\vec{j}\|dt = dt$. So

$$\begin{aligned}\int_C f\ ds &= \int_0^\pi f(\cos t, \sin t)\ dt \\ &= \int_0^\pi (\cos^2 t - \sin^2 t)\ dt \\ &= \int_0^\pi \cos 2t\ dt \\ &= 0.\end{aligned}$$

□

Example 13.1.2 *Let $f(x,y,z) = x^2 + z$. A curve C in $\mathbb{R}^3$ given by the equation $\vec{r}(t) = \cos t\vec{i} + \sin t\vec{j} + 2t\vec{k}$, for $0 \le t \le 2\pi$. Evaluate the line integral $\int_C f\ ds$.*

Solution. First we see that

$$ds = \|(-\sin t, \cos t, 2)\|\ dt = \sqrt{5}dt.$$

So the line integral is evaluated as follows:

$$\begin{aligned}\int_C f\ ds &= \int_0^{2\pi} (\cos^2 t + t)\sqrt{5}\ dt \\ &= \sqrt{5}\Big(\tfrac{1}{2} + \int_0^{2\pi} (\tfrac{1+\cos(2t)}{2}\ dt\Big) \\ &= \tfrac{\sqrt{5}}{2}(1 + 2\pi).\end{aligned}$$

□

13.1.2 Work, circulation, flux, etc.

Line integral is known to represent many concepts in classical mechanics and various other branches of science and technology. We will see them here as applications.

Vector field

If a strong magnet is placed somewhere, nearby iron particle at each location will feel the pulling force, which we understand it as a vector. This means that the magnet is manifested as an assignment of a vector at each point in the space. In other words, we are discussing a function(!) assigning a vector (that has direction and magnitude) to each point of the space. Such a function, that is vector-valued, defined on a certain region of the space (say $\mathbb{R}^3$, or even $\mathbb{R}^n$ for any n, for instance) is called a *vector field*, which takes the formula such as

$$\vec{F}(x_1, \ldots, x_n) = (f_1(\vec{x}), \ldots, f_\ell(\vec{x})).$$

We would like to point out that, for the greater generality, it is not necessary to restrict us to the special case $n = \ell$.

Work. Let us consider the case of $\mathbb{R}^2$ for a moment. Let C be a smooth curve, and let $\vec{r}(t) = x(t)\vec{i} + y(t)\vec{j}$, $a \le t \le b$, be its parametrization.

Suppose that a force/vector field $\vec{F} := (f_1(x, y), f_2(x, y))$ is given. Then the *work* done by $\vec{F}$ on C is known (to Physics) as

$$W = \int_C \vec{F}(x(t), y(t)) \cdot \vec{r}'(t)\ dt,$$

which is often written (making sense!) as

$$W = \int_C \vec{F} \cdot d\vec{r}.$$

This is also a line integral introduced before, because

$$\int_C \vec{F}(x(t), y(t)) \cdot \vec{r}'(t)\ dt = \int_C (\vec{F} \cdot \vec{T}) ds$$

where $\vec{T}(t) = \frac{\vec{r}'(t)}{\|\vec{r}'(t)\|}$. (Of course, to avoid obvious self-contradiction, we need restrict ourselves only to the curves with condition that the speed $\|\vec{r}'(t)\|$ of the motion/curve is nowhere zero.)

Circulation. Mathematical formulation of the concept called *circulation* of a vector field $\vec{F}$ along a curve $\vec{r}$ is exactly the same as that of *work*, above:

$$\int_C \vec{F} \cdot d\vec{r}.$$

But as the name/terminology suggests, there is a difference, that the curve/motion is simple-closed, meaning that the starting point of the curve coincides with the ending point (i.e., "closed"), and moreover, there are no other self-intersecting points in C (i.e., "simple").

Flux. It the curve C with parametrization $\vec{r}(t) = x(t)\vec{i} + y(t)\vec{j}$ with speed nowhere zero is drawn, it separates, at least locally, the plane into two sides. Then a vector field $\vec{F}$ defined on a region including the curve has the flow across the curve. A typical example is easier to picture in mind; if $\vec{F}$ represents the infinitesimal flow of a certain fluid (such as air, water, ...) on the plane, then the totality of infinitesimal flow across the curve C is called the flux of $\vec{F}$ across C. This is represented by

$$\int_C (\vec{F} \cdot N) ds$$

where N is the unit normal vector (field) along C. Now, notice that there are choices between two possibilities here. Namely,

$$N = \pm \frac{-y'(t)\vec{i} + x'(t)\vec{j}}{\| -y'(t)\vec{i} + x'(t)\vec{j}\|}.$$

13.1.3 Via the pullback-viewpoint

It is known that many line integrals can be understood as the pullback formalism. Such an interpretation provides not only conveniences but also helps us to understand wider variety of knowledge in compact, concise and better organized ways.

From here on, we fix the vector field $\vec{F}(x, y) = f_1(x, y)\vec{i} + f_2(x, y)\vec{j}$, and the curve C parametrized by $\vec{r}(t) = x(t)\vec{i} + y(t)\vec{j}$, $a \le t \le b$, with speed nowhere zero. (And, of course the ambient space is $\mathbb{R}^2$.)

Work and Circulation

Take the 1-form

$$\alpha := f_1(x, y)\ dx + f_2(x, y)\ dy.$$

Notice that $\vec{r}$ is a mapping from the interval $[a, b]$ in $\mathbb{R}$ to $\mathbb{R}^2$, and α is a 1-form on (part of) $\mathbb{R}^2$.

Then the work done by $\vec{F}$ to C is simply

$$W = \int_{[a,b]} \vec{r}^{\,*}\alpha.$$

So the following notation also makes sense (meaning is by the preceding formula):

$$W = \int_C f_1(x, y)\; dx + f_2(x, y)\; dy. \tag{13.3}$$

Flux

If we choose the unit normal vector $\vec{N} = \frac{-y'(t)\vec{\imath}+x'(t)\vec{\jmath}}{\sqrt{(x'(t))^2+(y'(t))^2}}$, then the flux Φ of $\vec{F}$ across C in the direction of $\vec{N}$ is equal to

$$\Phi = \int_{[a,b]} \vec{r}^{\,*}\beta$$

where $\beta = -f_1(x, y)\; dy + f_2(x, y)\; dx$. Thus the flux integral also receives the presentation

$$\Phi = \int_C -f_1(x, y)\; dy + f_2(x, y)\; dx. \tag{13.4}$$

Verification of all these follows by the construction above; we leave it to the student.

13.2 Path independence and Potential

The purpose of this section is to study some more on the line integrals.

13.2.1 Path independence

Let $p, q \in \mathbb{R}^n$ and C_1, C_2 be two curves (piecewise smooth) parametrized by $\vec{r}_k(t), (a_k \leq t \leq b_k)$ satisfying $\vec{r}_k(a_k) = p$ and $\vec{r}_k(b_k) = q$ for $k = 1, 2$, respectively. Then the values of line integrals

$$L_k = \int_{C_k} \vec{F} \cdot d\vec{r}, \ (k = 1, 2)$$

shall depend, naturally, on the choices of paths as well as the vector field $\vec{F}(x_1, \dots, x_n)$. But, as in Physics, if this line integral is applied to represent the concept of work done to the paths C_1, C_2 by the same force field $\vec{F}$, it is a natural question whether the amount of the work (i.e., the value of the line integral) stays invariant (therefore it is independent of the choices of paths and their parametrizations).

It has turned out to be not only a natural question in Physics but also a question with a significance in Mathematics. The problem concerns the following

> Let $p, q, \vec{r}_k, \vec{F}$ are given as in the beginning of this section. When does
>
> $$\int_{C_1} \vec{F} \cdot d\vec{r} = \int_{C_2} \vec{F} \cdot d\vec{r}$$
>
> hold?

It appears resonable to find the answer from the attributes of the vector field $\vec{F}$, and therefore ask when the values of the integrals depend only on the endpoints p, q of the paths.

If $\vec{F} = \nabla f$ for some real-valued (i.e., scalar-valued) function. then the chain rule has shown that

$$\begin{aligned} \vec{F}(\vec{r}(t)) \cdot \vec{r}'(t) &= \nabla f\big|_{\vec{r}(t)} \cdot \vec{r}'(t) \\ &= \frac{d}{dt} f(\vec{r}(t)). \end{aligned}$$

This leads us to:

Theorem 13.2.1 *If the vector field $\vec{F}(\vec{x})$ satisfies $\vec{F} = \nabla f$ for some real-valued function f defined on a region R in $\mathbb{R}^n$, then the identity*

$$\int_C \vec{F} \cdot d\vec{r} = f(q) - f(p)$$

holds for any path in R with initial point at p and terminal point at q.

This triggers another series of questions: When does such f exist? If it exists, how can one compute/find it? Is the existence of such f the only way of deducing the path-independence? And so forth. These questions naturally lead us to the following sections.

13.2.2 Potential

A vector field on $\mathbb{R}^n$ takes a coordinate expression such as

$$\vec{F}(x_1, \ldots, x_n) = (f_1(\vec{x}), \ldots, f_n(\vec{x})).$$

We are asking whether there is a function $u = u(x_1, \ldots, x_n)$ satisfying $\vec{F} = \nabla u$ on $\mathbb{R}^n$. In terms of equations we are asking to solve for u in

$$\begin{cases} \frac{\partial u}{\partial x_1} = f_1 & \cdots (1) \\ \quad\vdots & \\ \frac{\partial u}{\partial x_n} = f_n & \cdots (n). \end{cases} \tag{13.5}$$

Such an equation is called overdetermined, because one function u is required to satisfy n-equations posed by potentially n functions $f_1, \ldots, f_n$.

Some conditions are necessary. If a solution u existed (as a many times differentiable function), then it would satisfy

$$\begin{aligned} f_{i,x_j} = \frac{\partial f_i}{\partial x_j} &= \frac{\partial}{\partial x_j}\left(\frac{\partial u}{\partial x_i}\right) \\ &= \frac{\partial^2 u}{\partial x_j \partial x_i} \\ &= \frac{\partial^2 u}{\partial x_i \partial x_j} \\ &= \frac{\partial f_j}{\partial x_i} = f_{j,x_i}, \end{aligned}$$

for every $i, j \in \{1, \ldots, n\}$.

Note. For short-hand notation. we denote by $f_{i,x_j} = \frac{\partial f_i}{\partial x_j}$, $f_{i;x_j x_k} = \frac{\partial^2 f_i}{\partial x_k \partial x_j}$ and so forth, unless there are chances for confusion.

Theorem 13.2.2 *If $F(x_1, \ldots, x_n) = (f_1(\vec{x}), \ldots, f_n(\vec{x}))$ is a C^1 smooth vector field defined on $\mathbb{R}^n$ satisfies the condition*

$$f_{i,x_j} = f_{j,x_i}, \ \forall i, j \in \{1, \ldots, n\},$$

at every point of $\mathbb{R}^n$, then there is a C^2 smooth function u on $\mathbb{R}^n$ satisfying $\nabla u = \vec{F}$.

Proof. We shall use the mathematical induction on n. Namely, we shall prove the case $k = 2$ first, Then, assuming that the theorem is true for $k = n$ and then show that the theorem holds for $k = n + 1$.

Step 0: Set up. We shall number the steps, so that the whole arguments fit to the mathematical induction. Our present goal is (for every integer $n \geq 2$) to do the following:

$P(k)$: *To solve for u in*

$$\begin{cases} u_{x_1} = f_1 \\ \quad \vdots \\ u_{x_k} = f_k \end{cases}$$

under the condition, called the compatibility condition, that

$$f_{m,x_\ell} = f_{\ell,x_m}, \ \forall m, \ell \in \{1, \ldots, k\}.$$

Step 1: Verify $P(2)$. The problem now is to solve for u in

$$\begin{cases} u_{x_1} = f_1(x_1, x_2) \\ u_{x_2} = f_2(x_1, x_2) \end{cases}$$

under the condition that $f_{1,x_2} = f_{2,x_1}$. Integrating the second equation with respect to the x_2-variable, we get

$$u(x_1, x_2) = \int_0^{x_2} f_2 \, dx_2 + v(x_1) \tag{13.6}$$

for some $v(x_1)$, a smooth function of x_1-variable. Then

$$u_{x_1} - \int_0^{x_2} f_{2,x_1}\, dx_2 = v'(x_1)$$

which becomes

$$v'(x_1) = u_{x_1} - \int_0^{x_2} f_{1,x_2}\, dx_2.$$

Notice that the right hand side is a function independent of the x_2-variable. Therefore, integrating with respect to the x_1-variable, we see that $v(x_1)$ exists. Together with (13.6), we see that

$$u = \int_0^{x_2} f_2\, dx_2 + v(x_1)$$

is a solution, and hence $P(2)$ is solvable.

Step 2: Verify $P(n)$ assuming $P(n-1)$. The n-th equation yields

$$u(x_1, \ldots, x_n) = \int_0^{x_n} f_n(x_1, \ldots, x_n)\, dx_n + v(x_1, \ldots, x_{n-1}) \tag{13.7}$$

for some smooth function $v(x_1, \ldots, x_{n-1})$. Now, for any $\ell \in \{1, \ldots, n-1\}$, one sees that

$$\begin{aligned} v_{x_\ell} &= u_{x_\ell} - \int_0^{x_n} f_{n,x_\ell}\, dx_n \\ &= u_{x_\ell} - \int_0^{x_n} f_{\ell,x_n}\, dx_n \\ &= f_\ell(x_1, \ldots, x_{n-1}, 0). \end{aligned}$$

Therefore $v(x_1, \ldots, x_{n-1})$ satisfies

$$\begin{cases} v_{x_1} = f_1(x_1, \ldots, x_{n-1}, 0) \\ \quad \vdots \\ v_{x_{n-1}} = f_{n-1}(x_1, \ldots, x_{n-1}, 0) \end{cases}$$

and by the property of $f_1, \ldots, f_{n-1}$ the compatibility condition

$$\frac{\partial}{\partial x_m} f_\ell(x_1, \ldots, x_{n-1}, 0) = \frac{\partial}{\partial x_\ell} f_m(x_1, \ldots, x_{n-1}, 0)$$

holds for all $m, \ell \in \{1, \ldots, n-1\}$. Therefore, v is the solution of the problem $P(n-1)$, and hence it exists by the induction hypothesis. So by (13.7), u exists as a solution to $P(n)$. This completes the proof. □

Example 13.2.1 *Find $u(x, y)$ satisfying $du = (x^3 - xy)\ dx - \frac{1}{2}x^2\ dy$*

Solution. Let $P = x^3 - xy, \quad Q = -\frac{1}{2}x^2$. Then the compatibility condition

$$P_y = -x = Q_x$$

is checked by a direct computation. So on $\mathbb{R}^2$ the solution u should exist, by Theorem 13.2.2. (Also proof of the theorem tells us how to solve it.)

From $u_x = x^3 - xy$, we obtain

$$u(x, y) = \tfrac{x^4}{4} - \tfrac{x^2 y}{2} + v(y), \tag{13.8}$$

for some function v depending only on the y-variable. So

$$-\tfrac{1}{2}x^2 = u_y = -\tfrac{x^2}{2} + v'(y).$$

So $v'(y) = 0$, and hence $v = c$, a constant. Now from (13.8) it follows that

$$u(x, y) = \tfrac{x^4}{4} - \tfrac{x^2 y}{2} + c$$

for some constant c. □

Example 13.2.2 *Solve for u in*

$$\begin{cases} u_x = e^x \cos y + yz & (1) \\ u_y = xz - e^x \sin y & (2) \\ u_z = xy + z. & (3) \end{cases}$$

Solution. The compatibility condition is easily checked; detail is left to the student to verify.

Now, integrating the first equation with x-variable, we obtain

$$u = e^x \cos y + xyz + v(y, z), \tag{13.9}$$

for some function v independent of the x-variable. Now,

$$\begin{aligned} v_y &= u_y - (-e^x \sin y + xz) \quad \text{(by the preceding equation)} \\ &= 0 \quad \text{(by (2),)} \end{aligned}$$

and

$$\begin{aligned} v_z &= u_z - xy \quad \text{(by the preceding equation)} \\ &= z \quad \text{(by (3)).} \end{aligned}$$

Hence we obtain that v depends only on z and

$$v = \tfrac{z^2}{2} + C.$$

So with (13.9), we obtain

$$u(x, y, z) = e^x \cos y + xyz + \tfrac{z^2}{2} + C,$$

for some constant C. □

Remark 13.2.1 Theorem 13.2.2 was proved for $\mathbb{R}^n$. But it can also be proved (by the same proof) on an open cell (an n-dimensional box). On the other hand, it cannot be proved on the punctured plane $P := \mathbb{R}^2 \setminus \{(0,0)\}$. Here is the counterexample:

Notice that there is no differentiable function u on the punctured plane $\mathbb{R}^2 \setminus \{(0,0)\}$ satisfying

$$du = \frac{-y}{x^2 + y^2}\, dx + \frac{x}{x^2 + y^2}\, dy.$$

[Verify this fact, as the right hand side is the total differential of $\tan^{-1} \frac{y}{x}$. But this $\tan^{-1} \frac{y}{x}$ does not define a continuous function on the punctured plane. Still, the student has to prove that there is no $u\colon P \to \mathbb{R}$ with du coinciding with the right hand side at every point of P.

Definition 13.2.1 A vector field $F = (f_1, \ldots, f_n)$ defined on a region R in $\mathbb{R}^n$ is called *conservative* if it admits a potential, i.e., a scalar function $u\colon R \to \mathbb{R}$ with $\vec{F} = \nabla u$.

Notice that the components of the conservative vector field must satisfy the compatibility condition. But the converse is in general not true, i.e., the vector field may not be conservative, and yet, its components can satisfy the compatibility conditions.

13.3 Surface integral

13.3.1 Regular surface

The concept to be considered in this section is the integrals on a surface in $\mathbb{R}^3$ and, more generally, the integrals on a k dimensional surface in $\mathbb{R}^{k+\ell}$. But we shall focus more upon the former.

First we need a parametrization of a surface in $\mathbb{R}^3$. In a parametric form it is, locally,

$$\vec{r}(s,t) = x(s,t)\vec{i} + y(s,t)\vec{j} + z(s,t)\vec{k}$$

with $(s,t) \in R$ where R is a region in $\mathbb{R}^2$. Then the surface intuitively would be the image $S = \vec{r}(R)$ in $\mathbb{R}^3$. But for the purpose of Calculus, we will call the smooth function $\vec{r}(s,t)$ a *surface*, or more precisely a *parametrized surface.*

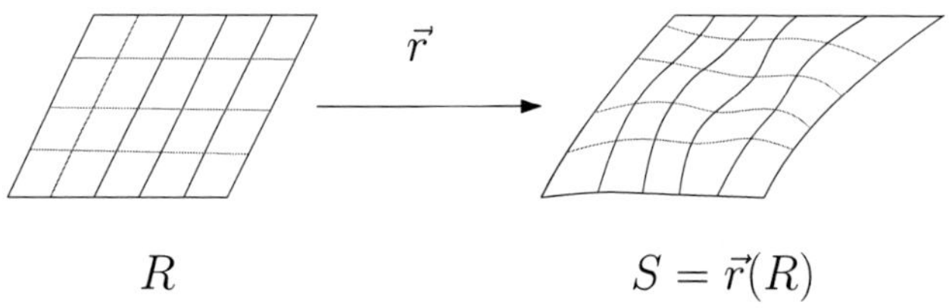

Figure 13.1: A parametrized surface

Notice that the small patch for S (i.e., a small parallelogram) generated by two vectors $\vec{r}_s, \vec{r}_t$ has the area

$$\vec{r}_s \times \vec{r}_t \cdot \vec{N}(\vec{r}(s,t))$$

where $\vec{N}(\vec{r}(s,t))$ is the unit vector at point $\vec{r}(s,t)$ in the "upward" direction of a vector perpendicular to the tangent plane of S. Hence the surface integral of f over S should be

$$\begin{aligned}\int_S f\ d\sigma_S &= \int_R f(\vec{r}(s,t))[\vec{r}_s \times \vec{r}_t \cdot \vec{N}(\vec{r}(s,t))]\ dsdt \\ &= \int_R f(\vec{r}(s,t))\|\vec{r}_s \times \vec{r}_t\|\ dsdt,\end{aligned}$$

because

$$\vec{N}(\vec{r}(s,t)) = \frac{\vec{r}_s \times \vec{r}_t}{\|\vec{r}_s \times \vec{r}_t\|}.$$

13.3.2 Definitions

In the setting above, we provide definition of some surface integrals.

Surface integral of a function

As mentioned above, the *surface element* for $S := \vec{r}(R)$ is given by

$$d\sigma_2 = \|\vec{r}_s \times \vec{r}_t\| \, ds\, dt.$$

and

$$\int_S f \; d\sigma_2 = \int_R f(\vec{r}(s,t))\|\vec{r}_s \times \vec{r}_t\| \; ds\, dt.$$

Flux of a vector field across the surface

As the surface analogue of flux across a curve, we shall define the flux integral across the surface $S = \vec{r}(R)$ given by the parametric equation $\vec{r}(s,t)$ as mentioned above. Of course we have to designate the normal vector (in fact, a normal vector field defined on the surface S); let us choose

$$\vec{N}(\vec{r}(s,t)) = \frac{\vec{r}_s \times \vec{r}_t}{\|\vec{r}_s \times \vec{r}_t\|}.$$

Then the *flux* of the vector field F across S in the direction of $\vec{N}$ should be

$$\Phi = \int_S \vec{F} \cdot \vec{N} \; d\sigma_2.$$

One can represent the flux integral via the language of pullback integral of differential forms. Notice that

$$\vec{F} \cdot \vec{N} \; \|\vec{r}_s \times \vec{r}_t\| = \vec{F} \cdot \vec{r}_s \times \vec{r}_t$$

and hence

$$\begin{aligned}\Phi &= \int_R \left(\vec{F}(\vec{r}(s,t)) \cdot \vec{r}_s \times \vec{r}_t\right) \; ds \; dt \\ &= \int_S \big(f_1(x,y,z)(y_s z_t - y_t z_s) + f_2(x,y,z)(z_s x_t - z_t x_s) \\ &\qquad\qquad + f_3(x,y,z)(x_s y_t - x_t y_s)\big) \; ds \; dt.\end{aligned}$$

This yields

$$\begin{aligned}\Phi &= \int_S f_1(x,y,z)\ dy \wedge dz + f_2(x,y,z)\ dz \wedge dx + f_3(x,y,z)\ dx \wedge dy \\ &= \int_R \vec{r}^{\,*}\big(f_1(x,y,z)\ dy \wedge dz + f_2(x,y,z)\ dz \wedge dx + f_3(x,y,z)\ dx \wedge dy\big).\end{aligned}$$

This also shows that any integration over a surface S of a differential 2-form defined on and about S can be realized as the flux of a certain vector field across the surface.

13.4 Applications of multiple integrals

Many applications of integrals on higher dimensions and, even on the nonflat spaces such as curves (line integral), surfaces (surface integral) and even for more general spaces (such as manifolds), are known. But we will be satisfied with a few traditional examples.

13.4.1 Mean value integral

Let R be an admissible region in $\mathbb{R}^n$ and f be a continuous function. Then both

$$\int_R f\ dV_n \text{ and } \int_R dV_n$$

exist. If $\int_R dV_n \neq 0$, then let

$$m := \frac{\int_R f\ dV_n}{\int_R dV_n}.$$

This m is called the *mean value* for the integral $\int_R f\ dV_n$ In summary we have the following version of mean-value theorem.

Theorem 13.4.1 *For any continuous function f defined on an admissible region R in $\mathbb{R}^n$ satisfying $\int_R dV_n \neq 0$, there exists a real number m such that*

$$\int_R f\ dV_n = \int_R m\ dV_n.$$

Notice that this average value m for f on the region R satisfies

$$\min_{x \in R} f(x) \leq m \leq \max_{x \in R} f(x).$$

13.4.2 Center of mass

The center of mass of an admissible region R is a point $\vec{p} = (p_1, \dots, p_n)$ which satisfies

$$\int_R (x_i - p_i)\ dV_n = 0, \quad \forall i = 1, \dots, n.$$

This implies that each p_i is the average value of the function x_i on the region R, i.e.,

$$p_i = \frac{\int_R x_i \, dV_n}{\int_R dV_n}, \quad \forall i.$$

There are several more examples such as the momentum in Classical Mechanics. One can also develop similar concepts for the lines, surfaces and so forth. But we choose not to include any further examples in these notes.

Chapter 14

Integral theorems

The Fundamental Theorem of Calculus (FTC) has a far-reaching important influence to the theory of functions. The main frame of this chapter is definitely based upon variations of FTC.

14.1 Green's theorem on a rectangular cell

Let C be the curve consisting of the sides of the cell E in $\mathbb{R}^2$ defined by

$$E = \{(x, y) \colon a_1 \le x \le a_2, b_1 \le y \le b_2\}.$$

Assume that C is oriented counterclockwise (ccw). Now let $\vec{F} = (f_1.f_2)$ be a smooth vector field defined on a region containing E inside. Then

$$\int_C \vec{F} \cdot d\vec{r} = \int_C f_1 \, dx + f_2 \, dy$$

as explained in (13.3). (This integral represents the work/circulation by the vector field $\vec{F}$ along the curve/path C, but we do not in particular like to be tied by such physical meaning.) Then

$$\begin{aligned}\int_C f_1 \, dx + f_2 \, dy \\ &= \int_{a_1}^{a_2} f_1(x, b_1) \, dx + \int_{a_2}^{a_1} f_1(x, b_2) \, dx \\ &\quad + \int_{b_1}^{b_2} f_2(a_2, y) \, dy + \int_{b_2}^{b_1} f_2(a_1, y) \, dy.\end{aligned}$$

This turns into

$$\begin{aligned}
&\int_C f_1\, dx + f_2\, dy \\
&\qquad = \int_{a_1}^{a_2} \big(f_1(x,b_1) - f_1(x,b_2)\big)\, dx + \int_{b_1}^{b_2} \big(f_2(a_2,y) - f_2(a_1,y)\big)\, dy \\
&\qquad = \int_{a_1}^{a_2} \Big(\int_{b_1}^{b_2} -f_{1,y}(x,y)\, dy\Big)\, dx + \int_{b_1}^{b_2} \Big(\int_{a_1}^{a_2} f_{2,x}(x,y)\, dx\Big)\, dy \\
&\qquad = \int_{a_1}^{a_2} dx \int_{b_1}^{b_2} \big(-f_{1,y}(x,y) + f_{2,x}(x,y)\big)\, dy \\
&\qquad = \int_E \big(-f_{1,y}(x,y) + f_{2,x}(x,y)\big) dV_2.
\end{aligned}$$

As the summary, we have

Theorem 14.1.1 (Green's theorem on a cell) *If a smooth vector field $\vec{F}(x,y) = f_1(x,y)\vec{i} + f_2(x,y)\vec{j}$ is defined on an open region containing the cell $E = \{(x,y)\colon a_1 \le x \le a_2, b_1 \le y \le b_2\}$, and if C is the boundary curve of E oriented counterclockwise, then*

$$\int_C f_1 dx + f_2 dy = \int_E \big(-f_{1,y} + f_{2,x}\big)\, dV_2.$$

This theorem can be extended. Consider a "body" B constructed as the union of finitely many cells. Then the overlapped cells can be subdivided so that the same body consists of cells with boundary edges intersected only or no intersections at all. Then the composition of two cells sharing edges will cause an interesting effect to the result of the preceding theorem in such a way that line integral along the common edges cancel whereas the double integral on two cells add up. This yields the following:

Proposition 14.1.1 (Green's theorem on a "body") *If a smooth vector field $\vec{F}(x,y) = f_1(x,y)\vec{i} + f_2(x,y)\vec{j}$ is defined on an open region containing the body B, and if ∂B denotes the boundary of B each of whose component curve is oriented counterclockwise, then*

$$\int_{\partial B} f_1 dx + f_2 dy = \int_B \big(-f_{1,y} + f_{2,x}\big)\, dV_2.$$

Now taking the limit as the bodies approximate the admissible regions, it finally leads us to the following

Theorem 14.1.2 (Green's theorem) *If a smooth vector field* $\vec{F}(x,y) = f_1(x,y)\vec{i} + f_2(x,y)\vec{j}$ *is defined on an open region containing an admissible region* R *with its boundary* ∂R *piecewise smooth and oriented counterclockwise, then*

$$\int_{\partial R} f_1\,dx + f_2\,dy = \int_R \big(-f_{1,y} + f_{2,x}\big)\,dV_2.$$

14.2 Green's theorem by differential forms

Recall the definition of differential k-forms. The convention is that by a 0-form we mean a smooth function.

At this stage, we introduce another well-known operation on differential forms called the *exterior differential* d as follows:

Definition 14.2.1 For each $k \in \{0, \dots, n\}$, the *exterior differential operator* $d(= d_k)$ is defined on the set $\Omega^k(D)$ of differential k-forms on an open set D in $\mathbb{R}^n$ with values in $\Omega^{k+1}(D)$ satisfying

(1) $d(f(x_1, \dots, x_n)\, dx_{i_1} \wedge \dots \wedge dx_{i_k}) = \sum_{\ell=1}^{n} \frac{\partial f}{\partial x_\ell}\, dx_\ell \wedge dx_{i_1} \wedge \dots \wedge dx_{i_k}$

(2) $d(\alpha + \beta) = d\alpha + d\beta$

for any smooth function $f(\vec{x})$ and any $\alpha, \beta \in \Omega^k(D)$. $d\alpha$ is called the *exterior derivative* of α.

Then several more properties are known for this exterior operator, From the definition one may deduce:

- $df = f_{x_1}\, dx_1 + \dots + f_{x_n}\, dx_n$.
- $d(d\alpha) = 0$ as a $(k+1)$-form.

There are more properties for d, but we shall not go into any further in this direction.

14.2.1 What the exterior differentiation can do

Let $\alpha = f_1(x_1, x_2)\, dx_1 + f_2(x_1, x_2)\, dx_2$ be defined on an open set D in $\mathbb{R}^2$ containing an admissible region R inside, with its boundary curve ∂R, piecewise smooth and oriented counterclockwise. Then

$$\begin{aligned} d\alpha &= f_{1,x_1}\, dx_1 \wedge dx_1 + f_{1,x_2}\, dx_2 \wedge dx_1 \\ &\quad + f_{2,x_1}\, dx_1 \wedge dx_2 + f_{2,x_2}\, dx_2 \wedge dx_2 \\ &= \big(- f1, x_2 + f_{2,x_1} \big)\, dx_1 \wedge dx_2 \end{aligned}$$

So classic Green's theorem in terms of exterior differentiation is stated as follows:

Theorem 14.2.1 (Green's theorem by differential forms) *If* $\alpha \in \Omega^1(D)$ *for an open set containing an admissible region* R *with piecewise smooth boundary* ∂R *oriented counterclockwise, then*

$$\int_{\partial R} \alpha = \int_R d\alpha.$$

Remark 14.2.1 As one would wonder, it is known that this version of Green's theorem has been extended to higher dimensions and several more general situations. We will not go into any deeper, but still, there is a merit in reformulating Green's theorem by differential forms. The reader will see that in the subsequent sections.

14.2.2 Some examples

Example 14.2.1 Let $C:\ \vec{r}(t) := (2\cos t, 3\sin t)$. Then for $0 \le t \le 2\pi$, this is a parametrization of an ellipse centered at $(0,0)$ with major and minor axes along the coordinate axes, respectively. Then with $R =$ the elliptic region bounded by C, we have

$$\int_C x\,dy = \int_R dx \wedge dy = \int_R dV_2.$$

Thus the area of R is equal to

$$\begin{aligned}\int_C x\,dy &= \int_0^{2\pi} 2\cos t\ d(3\sin t)\\ &= 6\int_0^{2\pi} \cos^2 t\ dt\\ &= 6\pi.\end{aligned}$$

Check also that

$$\begin{aligned}\text{Area }(R) &= \int_R dV_2 = \int_R dx \wedge dy\\ &= \int_{\partial R} x\ dy\\ &= \int_{\partial R} -y\ dx\\ &= \tfrac{1}{2}\int_{\partial R} x\ dy - y\ dx,\end{aligned}$$

for any compact region bounded by a piecewise smooth curve with finitely many singularities.

Example 14.2.2 (Exercise, in fact?) What happens to $\int_S dx_1 \wedge dx_2$ for $S = \{(x_1, x_2, x_3)\colon x_1^2 + x_2^2 + x_3^2 = 1, x_3 \geq 0\}$? It is still true that

$$\int_S dx_1 \wedge dx_2 = \int_{\partial S} x_1 \, dx_2.$$

But then the value of the line integral is π as ∂S is the unit circle centered at $(0, 0)$ (A direct computation is given in the preceding example!) But the area of S is 2π. Obviously, this shows that $\int_S dx_1 \wedge dx_2 \neq \int_S d\sigma_2$. It may be an interesting exercise for the student to explain what happens really.

14.3 Stokes' theorem on surfaces

Stokes's theorem on surface is a generalization of Green's theorem, treating "admissible" region on a smooth surface. Let S be a surface given by the parametric equation

$$\vec{X}(u_1,u_2) = x_1(u_1,u_2)\vec{i} + x_2(u_1,u_2)\vec{j} + x_3(u_1,u_2)$$

defined on an open set D in $\mathbb{R}^2_{\vec{u}}$. Assume also that $\vec{X}_{u_1} \times \vec{X}_{u_2}$ is nowhere zero. This defines a regular surface $S := \vec{X}(D)$.

Incidentally, any admissible region, say $\tilde{R}$, on S is constructed from an admissible region, say R in D with piecewise smooth boundary ∂R oriented counterclockwise by

$$\tilde{R} = \vec{X}(R).$$

In such a case, the normal vector field for S is defined to be

$$\vec{N} = \frac{\vec{n}}{\|\vec{n}\|},$$

where $\vec{n} = \vec{X}_{u_1} \times \vec{X}_{u_2}$. If $\vec{r}(t)$ is the parametrization of ∂R, then the boundary $\partial\tilde{R}$ is parametrized by $\vec{X} \circ \vec{r}(t) := \vec{X}(\vec{r}(t))$.

The Curl of a vector field

At this point we shall introduce another well-known operator called the *curl* of a vector field. If the vector field is given by

$$\vec{F}(\vec{x}) = f_1(\vec{x})\vec{i} + f_2(\vec{x})\vec{j} + f_3(\vec{x})\vec{k},$$

then its curl is defined to be

$$\mathit{curl}\ \vec{F} = (f_{3,x_2} - f_{2,x_3})\vec{i} - (f_{1,x_3} - f_{3,x_1})\vec{j} + (f_{2,x_1} - f_{1,x_2})\vec{k}.$$

This last can be represented by the determinant form

$$\begin{vmatrix} \vec{i} & \vec{j} & \vec{k} \\ \frac{\partial}{\partial x_1} & \frac{\partial}{\partial x_2} & \frac{\partial}{\partial x_3} \\ f_1 & f_2 & f_3 \end{vmatrix}.$$

Hence the expression $\nabla \times \vec{F}$, via the cross product notation, is justified to represent *curl* $\vec{F}$.

Aside: In case the vector field $\vec{F}$ is defined on the xy-plane, we shall regard it as $\vec{F} = \big(f_1(x,y,0), f_2(x,y,0), 0\big)$ and hence its *curl* is defined by

$$curl\ \vec{F} = (0, 0, f_{2,x} - f_{1,y}).$$

Therefore the compatibility condition for $\vec{F}$ to admit a potential f satisfying $\nabla f = \vec{F}$ is the same as the condition

$$curl\ \vec{F} = \vec{0}.$$

Classic Stokes' theorem

Let the surface S, its parametrization $\vec{X}(u_1, u_2)$, the surface region $\tilde{R} \subset S$, the region $R \in \mathbb{R}^2_{\vec{u}}$ which $\tilde{R}$ is built upon, and the unit normal vector fields $\vec{N}$ are as above (in the beginning of this section). Then with the notation that $\vec{e}_1 = \vec{i}, \vec{e}_2 = \vec{j}, \vec{e}_3 = \vec{k}$, we present the following

Theorem 14.3.1 (Classic Stokes' theorem) *If $\vec{F}(x_1, x_2, x_3) = \sum_{\ell=1}^{3} f_\ell(\vec{x})\, \vec{e}_\ell$ is a smooth vector field in an open region in $\mathbb{R}^3$ containing the surface S, then*

$$\int_{\partial\tilde{R}} \vec{F} \cdot d\tilde{r} = \int_{\tilde{R}} (\nabla \times \vec{F}) \cdot \vec{N}\ d\sigma_2,$$

where $\tilde{r}(t)$ is the parametrization of the boundary curve $\partial\tilde{R}$ in $\mathbb{R}^3$.

Notice that, in case the surface S is the xy-plane $\mathbb{R}^2$ in the xyz-space $\mathbb{R}^3$, this version of Stokes's theorem is exactly the same as Green's theorem.

Proof of Classic Stokes' theorem (Optional). Let $\vec{r}(t) = u_1(t)\vec{i} + u_2(t)\vec{j}$ be a parametrization of the boundary curve ∂R of the admissible region R in $\mathbb{R}^2_{\vec{u}}$ so that $\vec{X}(u_1(t), u_2(t))$ becomes a parametrization of $\partial\tilde{R}$, the boundary curve of the surface region $\tilde{R}$ in S. Also let

$$\beta := \sum_{\ell=1}^{3} f_\ell(x_1, x_2, x_3)\ dx_\ell.$$

Then by Theorem 12.4.4 we see that

$$\int_{\partial R} \vec{X}^*(\beta) = \int_{\partial\tilde{R}} \beta.$$

Notice that $\int_{\partial\tilde{R}}\beta = \int_{\partial\tilde{R}}\vec{F}\cdot d\tilde{r}$, and hence

$$\int_{\partial R}\vec{X}^*(\beta) = \int_{\partial\tilde{R}}\beta = \int_{\partial\tilde{R}}\vec{F}\cdot d\tilde{r}. \tag{14.1}$$

On the other hand, it also holds that

$$\int_R \vec{X}^*(d\beta) = \int_{\tilde{R}} d\beta.$$

by computation with the exterior differential operator d. Moreover, one sees by a direct calculation that

$$\vec{X}^*(d\beta) = d\big(\vec{X}^*\beta\big).$$

Therefore,

$$\begin{aligned}\int_R \vec{X}^*(d\beta) &= \int_R d\big(\vec{X}^*\beta\big)\\ &= \int_{\partial R}\vec{X}^*\beta \quad \text{(by Theorem 14.2.1)}\\ &= \int_{\partial\tilde{R}}\beta\\ &= \int_{\partial\tilde{R}}\vec{F}\cdot d\tilde{r} \quad \text{(by (14.1))}.\end{aligned}$$

On the other hand,

$$\begin{aligned}d\beta &= \sum_{\ell=1}^{3}\sum_{m-1}^{3}\frac{\partial f_\ell}{\partial x_m}\,dx_m\wedge dx_\ell\\ &= (f_{3,x_2}-f_{2,x_3})\,dx_2\wedge dx_3-(f_{1,x_3}-f_{3,x_1})\,dx_3\wedge dx_1\\ &\qquad +(f_{2,x_1}-f_{1,x_2})\,dx_1\wedge dx_2.\end{aligned}$$

Then one can verify that

$$\int_{\tilde{R}} d\beta = \int_{\tilde{R}}(\nabla\vec{F})\cdot\vec{N}\,d\sigma_2.$$

Altogether, we obtain

$$\int_{\partial\tilde{R}}\vec{F}\cdot d\tilde{r} = \int_{\tilde{R}}(\nabla\times\vec{F})\cdot\vec{N}\,d\sigma_2.$$

This completes the proof. □

Remark 14.3.1 This last step of the proof of Stokes' theorem states the theorem in terms of the differential forms:

$$\int_{\partial \tilde{R}} \beta = \int_{\tilde{R}} d\beta.$$

This is strongly analogous to Green's theorem in the differential form language. This version works in more general cases with higher degree differential forms with spaces more general than surfaces, but we do not include any further exposition here.

It will be a great exercise to carry out symbolic calculations to verify the equalities in the proof above. We leave them to the student.

Remark 14.3.2 (Classic Green's theorem, again) For a region R surrounded by its boundary ∂R piecewise smooth with finitely many singularities, it holds that

$$\int_{\partial R} \vec{F} \cdot \vec{T}\, ds = \int_{R} (\nabla \times \vec{F}) \cdot \vec{k}\, dV_2.$$

Here, of course $F(x, y, z) = f_1(x, y, 0)\vec{i} + f_2(x, y, 0)\vec{j} + 0\vec{k}$ and $\vec{N} = \vec{k}$.

14.4 The Divergence theorem

We start from the concept of divergence in Classical mechanics.

14.4.1 The divergence in $\mathbb{R}^n_{\vec{x}}$

In $\mathbb{R}^n$ equipped with the Cartesian coordinate system $(x_1, \ldots, x_n)$ and the dot product, the *divergence* Div $\vec{F}$ of a vector field $\vec{F} = \sum_{i=1}^n f_i(\vec{x})\vec{e}_i$ is defined to be

$$\text{Div } \vec{F} := \frac{\partial f_1}{\partial x_1} + \ldots + \frac{\partial f_n}{\partial x_n}.$$

Physicists likes to use the notation, in this regard, as follows:

$$\text{Div } \vec{F} = \nabla \cdot \vec{F}$$

which means the definition above.[1]

14.4.2 The divergence theorem for the region in $\mathbb{R}^2_{\vec{x}}$

The simplest case of the divergence theorem can be for an n-dimensional cell

$$E^n(A, B) = \{(x_1, \ldots, x_n) \colon a_i \leq x_u \leq b_i, \forall i = 1, \ldots, n\}$$

described by $A := (a_1, \ldots, a_n)$ and $B := (b_1, \ldots, b_n)$ which are elements of $\mathbb{R}^n$ satisfying the property $a_i < b_i$ for every $i \in \{1, \ldots, n\}$.

Theorem 14.4.1 *For any smooth vector field $\vec{F}$ defined on an open region containing $E := E^n(A, B)$, it holds that*

$$\int_{\partial E} \vec{F} \cdot \vec{N} \, d\sigma_{n-1} = \int_E Div \, \vec{F} \, dV_n,$$

where $\vec{N}$ is the outward unit normal vector to the boundary hypersurface ∂E.

[1]This definition as well as those for the gradient and the curl depend upon the choice of the coordinate system (in this version of Calculus, the Cartesian coordinate system has been taken), which has no mathematically justifiable reason to be preferred. For more fundamental definition, however, requires a deeper understanding—and hence will not be mentioned here.

Of course, there is an ambiguity on the formal level concerning how to interprete $\vec{N}$ at points of the "edge" consisting of the intersections of the flat boundary components. But that does not matter, as the $n-1$ dimensional hypersurface measure of the edge is 0 as they are congruent to a region in $\mathbb{R}^k$ for some $k < n-1$. Such a set (in this case the edge) is said to be *of measure zero.*

Once this is established, the student can see that the following theorem holds:

Theorem 14.4.2 (The divergence theorem) *Let R be an admissible compact region with piecewise smooth boundary ∂R in $\mathbb{R}^n$, and let $\vec{F}$ be a smooth vector field defined on an open region containing R, then*

$$\int_{\partial R} \vec{F} \cdot \vec{N}\ d\sigma_{n-1} = \int_R Div\ \vec{F}\ dV_n,$$

where $\vec{N}$ is the outward unit normal vector to the boundary hypersurface ∂R.

Again, the region R being admissible includes the property that the set of points at which S is not smooth is of measure zero so that the integral on the boundary hypersurface ∂R in the statement of Theorem makes sense.

14.4.3 Proof in case for a cell E in $\mathbb{R}^3_{\vec{x}}$.

The outward flux integral Φ for $E = E^3(A, B)$ can be explicitly written. Obviously the boundary ∂E consists of six faces with matching outward normal vectors N, respectively:

$$\begin{array}{ll}
M_{1+} = \{(b_1, x_2, x_3)\colon a_k \le x_k \le b_k, \forall k \neq 1\} & \vec{N} = \vec{e}_1 = (1,0,0) \\
M_{1-} = \{(a_1, x_2, x_3)\colon a_k \le x_k \le b_k, \forall k \neq 1\} & \vec{N} = -\vec{e}_1 = (-1,0,0) \\
M_{2+} = \{(x_1, b_2, x_3)\colon a_k \le x_k \le b_k, \forall k \neq 2\} & \vec{N} = \vec{e}_2 = (0,1,0) \\
M_{2-} = \{(x_1, a_2, x_3)\colon a_k \le x_k \le b_k, \forall k \neq 2\} & \vec{N} = -\vec{e}_2 = (0,-1,0) \\
M_{3+} = \{(x_1, x_2, b_3)\colon a_k \le x_k \le b_k, \forall k \neq 3\} & \vec{N} = \vec{e}_3 = (0,0,1) \\
M_{3-} = \{(x_1, x_2, a_3)\colon a_k \le x_k \le b_k, \forall k \neq 3\} & \vec{N} = -\vec{e}_3 = (0,0,-1).
\end{array}$$

So the outward flux Φ of $\vec{F} = (f_1, f_2, f_3)$ is

$$\Phi = \sum_{i=1}^{3} \Big(\int_{M_{i+}} \vec{F} \cdot \vec{N}\ d\sigma_2 - \int_{M_{i-}} \vec{F} \cdot \vec{N}\ d\sigma_2 \Big).$$

Then for instance

$$\begin{aligned}\int_{M_{1+}} \vec{F}\cdot\vec{N}\ d\sigma_2 - \int_{M_{2-}} \vec{F}\cdot\vec{N}\ d\sigma_2 \\ &= \int_{E_1} \big(f_1(b_1,x_2,x_3) - f_1(a_1,x_2,x_3)\big)\ dV_2 \\ &= \int_E \frac{\partial f_1}{\partial x_1}\ dV_3,\end{aligned}$$

where $E_1 = E^2((a_2,a_3),(b_2,b_3)) = \{(x_2,x_3)\colon a_k \le y \le b_k,\ k=2,3\}$.

Repeating this with $M_{j\pm}$ for $j=2,3$ and summing all up, one obtains

$$\Phi = \int_E \text{Div}\ \vec{F}\ dV_3.$$

This completes the proof. □

14.4.4 By way of differential form formalism

One can rewrite the proof in the preceding section in the language of differential forms and wedge products. It is not a new proof, but rather, is a different way of writing the same thing.

$$\begin{aligned}\int_{\partial R} (\vec{F}\cdot\vec{N})\ d\sigma_2 &= \int_{\partial R} f_1\ dx_2\wedge dx_3 + f_2\ dx_3\wedge dx_1 + f_3\ dx_2\wedge dx_1 \\ &= \int_R d\big(f_1\ dx_2\wedge dx_3 + f_2\ dx_3\wedge dx_1 + f_3\ dx_2\wedge dx_1\big) \\ &= \int_R (f_{1,x_1} + f_{2,x_2} + f_{3,x_3})\ dx_1\wedge dx_2\wedge dx_3 \\ &= \int_R \text{Div}\ \vec{F}\ dV_3.\end{aligned}$$

This may be something in which one can see the use of symbolic algebra. And of course in this regard, the divergence theorem is another generalization of Green's theorem for flux integral. (Notice that the 2nd identity is a formal analogue of Green's theorem in the language of differential forms and wedge product.)

Chapter 15

Exams of Semester 2019-02

15.1 Examination 1

Exam 1 (10%) — Math 102
Calculus II/POSTECH
8:00-9:00 P. M. Thu., 26-Sept.-2019

Directions and notes: (1) No calculators are allowed.

(2) Grading applies only on what are written on the exam paper. No partial credits are allowed essentially.

(3) This exam consists of 4 problems; each problem is worth 2.5 points. The decimal values will be kept undiscarded until the total score of all 4 examinations are added.

[**1**] Let $\mathbf{A} = (1, a, -1), \mathbf{B} = (0, 0, b), \mathbf{C} = (0, 2, -2)$ be three points in the xyz-space (i.e., $\mathbb{R}^3$). Find values for a, b such that there is a unique plane passing through these 3 points and that the plane is perpendicular to the vector $\vec{Q} := \vec{i} + \vec{j} + \vec{k}$.

[**2**] Find the linearization (or, equivalently, an equation of the tangent plane) at $(0, 0)$ to the graph of the function $f(x, y) = x \cos y - ye^x$.

[3] Show that $f(x, y) = 1 + x + y + xy^2$ is differentiable at $(1, -1)$.

[4] Let $\vec{v} = (a, b)$ be a unit vector. Assume that a two-variable function $f(x, y)$ admits all mixed partial derivatives of order up to and including 2, satisfying

$$\left.\frac{\partial f}{\partial x}\right|_{(0,0)} = a, \quad \left.\frac{\partial f}{\partial y}\right|_{(0,0)} = b$$
$$\left.\frac{\partial^2 f}{\partial x^2}\right|_{(0,0)} = 2 = \left.\frac{\partial^2 f}{\partial y^2}\right|_{(0,0)}, \quad \left.\frac{\partial^2 f}{\partial x \partial y}\right|_{(0,0)} = 0 = \left.\frac{\partial^2 f}{\partial y \partial x}\right|_{(0,0)}.$$

Define by $g(t) = f(t\vec{v})$. Find the values of $g'(0)$ and $g''(0)$.

15.2 Midterm Examination

Midterm Exam (30%) — Math 102

Calculus II/POSTECH

6:00-9:00 p. m., Mon., 21-October-2019

Directions and notices:

(1) Exam does not allow possession of any calculators, computers or mobile phones. Keeping them in the bags is allowed, but take them out is considered as an illegal action.

(2) Grading applies only on what are written on the exam paper. No partial credits are allowed essentially.

(3) This exam consists of 10 problems; each problem is worth 3 points.

[**1**] Let $f(x, y) = x^3 + x^2 + xy + 3y^2$. Show that $f(0, 0)$ is a local minimum.

[**2**] Let P_1 be the tangent plane to the surface $z = x\cos y - ye^x$ at $(0, 0, 0)$, and let P_2 the plane given by the equation $2x+y-2z = 5$. Find a parametric equation for the line that coincides with the intersection $P_1 \cap P_2$.

[**3**] Let

$$f(x, y) = \begin{cases} \dfrac{x^2 y}{x^4 + y^2} & \text{if } (x, y) \neq (0, 0), \\ 0 & \text{if } (x, y) = (0, 0). \end{cases}$$

Does $\left[\frac{\partial}{\partial x}\left(\frac{\partial f}{\partial y}\right)\right]_{(0,0)}$ exist? If so, find its value. If not, say so with justification.

[**4**] Let $\vec{v} = (a, b)$ be a unit vector. Assume that a function $f(x, y)$ is of class C^2, and satisfies

$$\left.\frac{\partial f}{\partial x}\right|_{(0,0)} = a, \quad \left.\frac{\partial f}{\partial y}\right|_{(0,0)} = b$$

$$\left.\frac{\partial^2 f}{\partial x^2}\right|_{(0,0)} = 2 = \left.\frac{\partial^2 f}{\partial y^2}\right|_{(0,0)}, \quad \left.\frac{\partial^2 f}{\partial x \partial y}\right|_{(0,0)} = 0 = \left.\frac{\partial^2 f}{\partial y \partial x}\right|_{(0,0)}.$$

Define by $g(t) = f(t\vec{v})$. Find the value of $g''(0)$.

[5] Define $f(x, y, z) := \sqrt[3]{xyz}$. Let $\vec{u} = (a_1, a_2, a_3)$ be a unit vector. Show that the directional derivative satisfies $\mathrm{Dir}_{\vec{u}} f(0, 0, 0) = \sqrt[3]{a_1 a_2 a_3}$.

[6] The surface of a mountain is described as the graph of

$$f(x, y) = \sin(\pi x) \cdot \cos(\pi y),$$

where the x-axis points towards East and the y-axis points towards North. Now a climber located at the point $(0, 1, 0)$ is moving along towards North-west. Find the slope which the climber feels in that direction.

[*Note*: Slope may take a negative value if the climber is going down.]

[7] A delivery company accepts only rectangular boxes the sum of whose length and girth (perimeter of a cross-section) does not exceed 108 inches. Find the dimensions (i.e., length, depth and height) of an acceptable box of the largest volume.

[8] Let Ω denote the set on the xy-plane ($\mathbb{R}^2$) defined by the inequalities

$$0 \le x \le 2, \ y \ge 0, \ y \le x + 1, \ y \le -x + 3.$$

Find the maximum and the minimum of the function $f(x, y) = 10x - 5y$ defined on the set Ω.

[9] Four points $(-1, -1), (0, 1), (1, 2), (2, 2)$ are given. Find an equation of the straight line which minimizes the sum of squares of the horizontal deviations from the points (i.e., the x-regression line).

[10] Let G be the curve in the xy-plane (equipped with xy-coordinates) described by the equation $r = 1 + \cos\theta$ (a "cardioid") in the polar coordinate system (r, θ). Find an equation, in xy-coordinates, of the tangent line at $\theta = \frac{\pi}{3}$.

15.3 Examination 3

Exam 3 (10%) — Math 102
Calculus II/POSTECH
8:00-9:00 P. M. Thu., 21-Nov.-2019

Please Note: (1) No calculators are allowed.
(2) Grading applies only on what are written on the exam paper.
(3) This exam consists of 4 problems each of which is worth 2.5 points.

[**1**] Evaluate the iterated integral

$$A = \int_0^2 \Big(\int_x^2 \frac{1}{y^2+1}\, dy \Big)\, dx. \tag{15.1}$$

[**2**] Find the volume of the solid region D in $\mathbb{R}^3$ defined by the inequalities $x^2+y^2+z^2 \le 2z$ and $x^2+y^2+z^2 \le 1$.

[**3**] Let C be the curve tracing the upper half of the unit circle from $(1,0)$ to $(-1,0)$, along the counterclockwise direction, in the xy-plane $\mathbb{R}^2$. The vector field $\vec{F}$ is given by $\vec{F}(x,y) = \Big(\frac{-y}{\sqrt{x^2+y^2}}, \frac{x}{\sqrt{x^2+y^2}} \Big)$ on $\mathbb{R}^2 - \{(0,0)\}$. Evaluate the line integral $W = \int_C \vec{F} \cdot \vec{T} ds$ which represents the work done to C by $\vec{F}$.

[**4**] A curve C in $\mathbb{R}^2$ is given in such a way that it starts at the point $P_1 = (1,0)$, travels along the quarter-circle on the xy-plane to $P_2 = (0,1)$, and then travels further along a straight line segment to the terminal point $P_3 = (3,0)$. Evaluate the line integral

$$\int_C \big((y+e^x)\, dx + (x + 3\sin y)\, dy \big).$$

15.4 Final Examination

Math 102—Final Examination (50%)
Multivariable Calculus, Semester 2019-02, POSTECH
6:00-9:00 p.m. Tuesday, 17 December 2019

Directions and notes: (1) No electronic devices are allowed.

(2) Grading applies only on what are written on the exam paper. No partial credits are allowed essentially.

(3) This exam consists of 10 problems, each of which is worth 5 points.

[**1**] Let $n \geq 3$ be an odd number. Define the function f on $\mathbb{R}^n$ by

$$f(x_1, \ldots, x_n) := \sqrt[n]{x_1 \cdots x_n}.$$

Let $\vec{u} = (a_1, \ldots, a_n) \in \mathbb{R}^n$ be a unit vector. Show that the directional derivative satisfies

$$\mathrm{Dir}_{\vec{u}} f(0, \ldots, 0) = \sqrt[n]{a_1 \cdots a_n}.$$

[**2**] Evaluate the integral $\displaystyle\int_E \frac{x^2}{x^2+y^2}\, d^3X$, where E is the solid region bounded by the surfaces described by $z = x^2 + 2y^2$ and $z = 36 - 3x^2 - 2y^2$.

[**3**] Let E be the intersection curve of the plane $x + y + 2z = 2$ and the paraboloid $z = x^2 + y^2$. Find the points on the set E that are the closest to the origin.

[**4**] Evaluate the iterated integral

$$A := \int_0^2 \left(\int_0^1 \left(\int_x^2 z e^{y^2+1}\, dy \right) dz \right) dx.$$

[**5**] Let C be the curve in $\mathbb{R}^2$, parametrized by the equation

$$\vec{X}(t) = (\sin 2t)\,\vec{i} + (\sin t)\,\vec{j}$$

with $0 \leq t \leq \pi$. Denote by D the region surrounded by the curve C. Find the area of D.

[6] Let S be the surface in $\mathbb{R}^3$ obtained by rotating the circle $\{(x, y, z) \in \mathbb{R}^3 \,|\, (x-3)^2 + y^2 = 1,\ z = 0\}$ about the y-axis. This S, traditionally called a *torus*, can be parametrized by

$$\vec{X}(u, v) = ((3 + \cos u)\cos v)\ \vec{i} + (\sin u)\ \vec{j} + ((3 + \cos u)\sin v)\ \vec{k},$$

with $-\pi \leq u \leq \pi$ and $-\pi \leq v \leq \pi$. (The student need not justify this fact). Also the vector field $\vec{F}$ is given by

$$\vec{F}(x, y, z) := (x + z)\vec{i} + (x + y^3)\vec{j} + (y - 2z)\vec{k}.$$

Now, evaluate the flux integral

$$\Phi := \int_S (\text{curl } \vec{F}) \cdot \vec{N}\, d\sigma, \quad \text{where} \quad \vec{N} = \frac{\vec{X}_u \times \vec{X}_v}{\|\vec{X}_u \times \vec{X}_v\|}.$$

[7] The vector field $\vec{F}$ is given by

$$\vec{F} = yz\ \vec{i} - xz\ \vec{j} + (\tan^{-1} z)\ \vec{k}.$$

Moreover, the surface S is given by the conditions

$$x^2 + y^2 + z^2 = 9, \quad 0 \leq z \leq 2.$$

(The surface S is the portion of the sphere of radius 3 centered at the origin, cut by two planes $z = 0$ and $z = 2$. Only the part between those two planes are taken to be S. See the picture below.)

Now, evaluate the **outward flux**, say Φ, (from the viewpoint of the sphere) of the vector field $\nabla \times \vec{F}$ (i.e., the curl of $\vec{F}$) across S.

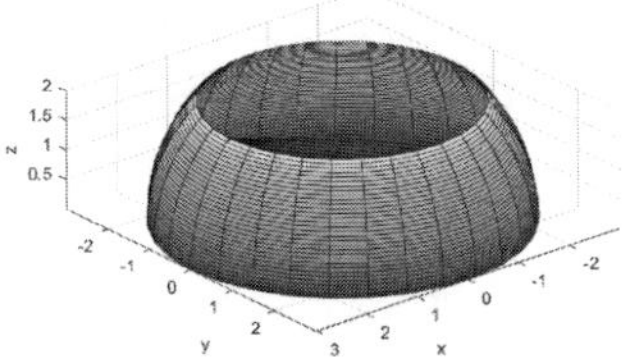

[8] The vector field $\vec{F}$ on the xy-plane is given by

$$\vec{F}(x,y) := \frac{-y+2}{(x-1)^2+(y-2)^2}\,\vec{i} + \frac{x-1}{(x-1)^2+(y-2)^2}\,\vec{j}.$$

Let C_R be the circle of radious $R > 0$ centered at the point $(1,2)$ oriented counterclockwise. Then, evaluate the work integral $\int_{C_R} \vec{F}\cdot\vec{T}\,ds$.

[9] A surface in $\mathbb{R}^3$ is defined by $S = \{(x,y,z) \in \mathbb{R}^3 : x+y-z=1, 0 \le x \le 1, 0 \le y \le 1\}$. The vector field $\vec{F}$ is given by $\vec{F}(x,y,z) = y\vec{i} + xy\vec{j} + z\vec{k}$. Finally, C denotes the boundary of S oriented consistently with the upward normal vector, according to the right hand rule. Then evaluate the work integral $W = \int_C \vec{F}\cdot\vec{T}\,ds$. [A vector $\vec{v}$ in $\mathbb{R}^3$ is said to be *upward* if $\vec{v}\cdot\vec{k} > 0$.]

[10] Let S be the side surface of a pyramid with apex at $(0,0,1)$ consisting of four triangles as in the picture below defined by the planes

$$x+z=1, \quad -x+z=1, \quad y+z=1, \quad -y+z=1$$

that are cut by the neighboring surfaces as well as by the planes $z=0$ and $z=1$. Note, however, that the bottom square is not part of S. Define the vector field $\vec{F}$ by

$$\vec{F}(x,y,z) = (2x+y^2)\vec{i} - (y+2)\vec{j} - (2+z)\vec{k}.$$

And, let $\vec{N}$ be the unit normal vector field to S with the $\vec{k}$ component positive.

Now, evaluate the flux integral $\Phi := \int_S \vec{F}\cdot\vec{N}\,d\sigma$.

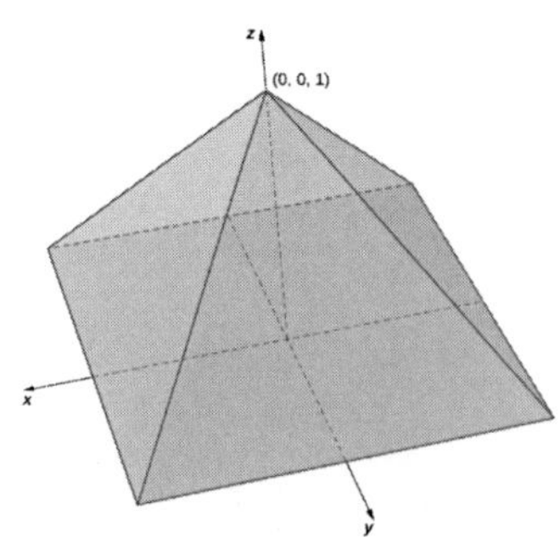

Index